T0213002

SpringerBriefs in Molecular Science

Biobased Polymers

Series editor

Patrick Navard, Sophia Antipolis cedex, France

Published under the auspices of EPNOE*_Springerbriefs in Biobased polymers_ covers all aspects of biobased polymer science, from the basis of this field starting from the living species in which they are synthetized (such as genetics, agronomy, plant biology) to the many applications they are used in (such as food, feed, engineering, construction, health, …) through to isolation and characterization, biosynthesis, biodegradation, chemical modifications, physical, chemical, mechanical and structural characterizations or biomimetic applications. All biobased polymers in all application sectors are welcome, either those produced in living species (like polysaccharides, proteins, lignin, …) or those that are rebuilt by chemists as in the case of many bioplastics.

Under the editorship of Patrick Navard and a panel of experts, the series will include contributions from many of the world's most authoritative biobased polymer scientists and professionals. Readers will gain an understanding of how given biobased polymers are made and what they can be used for. They will also be able to widen their knowledge and find new opportunities due to the multidisciplinary contributions.

This series is aimed at advanced undergraduates, academic and industrial researchers and professionals studying or using biobased polymers. Each brief will bear a general introduction enabling any reader to understand its topic.

*_EPNOE The European Polysaccharide Network of Excellence (www.epnoe.eu) is a research and education network connecting academic, research institutions and companies focusing on polysaccharides and polysaccharide-related research and business._

More information about this series at http://www.springer.com/series/15056

Gustav Sandin · Greg M. Peters
Magdalena Svanström

Life Cycle Assessment of Forest Products

Challenges and Solutions

 Springer

Gustav Sandin
SP Sustainable Built Environment
SP Technical Research Institute of Sweden
Gothenburg
Sweden

Magdalena Svanström
Chemistry and Chemical Engineering
Chalmers University of Technology
Gothenburg
Sweden

Greg M. Peters
Chemistry and Chemical Engineering
Chalmers University of Technology
Gothenburg
Sweden

ISSN 2191-5407   ISSN 2191-5415   (electronic)
SpringerBriefs in Molecular Science
ISBN 978-3-319-44026-2   ISBN 978-3-319-44027-9   (eBook)
DOI 10.1007/978-3-319-44027-9

Library of Congress Control Number: 2016948769

Printed on acid-free paper

This Springer imprint is published by Springer Nature
The registered company is Springer International Publishing AG
The registered company address is: Gewerbestrasse 11, 6330 Cham, Switzerland

Preface

Reducing environmental degradation and our dependency of finite resources are important motivations for developing a more bio-based society. In such a society, the most abundant renewable resource on the planet—forest biomass—will play a much more prominent role than in our current fossil-based society. To guide this transformation and obtain the potential environmental benefits of a more bio-based society, there is a need for high-quality, context-adapted environmental assessments.

Different types of environmental assessments are needed for decision-making concerned with different types of entities: sites, products, organisations, industry sectors, regions, nations, etc. For studies of products and services, life cycle assessment (LCA) is the most commonly used assessment tool worldwide. LCA is capable of assessing a wide range of environmental impacts over the entire life cycle of products and services, from resource[1] extraction (the "cradle"), via production, transportation and use, to waste management (the "grave").

Although there is an array of useful consensus documents guiding the LCA practitioner—the 14040/14044 International Organization for Standardization's (ISO) standard, the EN 16760 standard, the international reference life cycle data system (ILCD) handbook, the product environmental footprint (PEF) guide, to name a few—it can be rather challenging to carry out an LCA. Key challenges include the modelling of the product system and its interaction with the environment, the translation of emissions and resource use into quantified environmental impacts, and the interpretation and use of the results in various contexts. For example, how should one allocate environmental impacts between the many outputs of a biorefinery? How can one assess the climate, biodiversity, and water cycle impacts of forestry operations? How can one get the most out of LCA in research

---

[1]Resource acquisition is sometimes called raw material acquisition, but according to the European standard EN 16760:2015, it is important to distinguish between resources and raw materials in LCA studies involving biomass. Resources are flows without previous human transformation entering the system, whereas raw materials are intermediary flows within the product system (Swedish Standards Institute (SIS) 2015).

and development projects? What do LCA results say in relation to the global challenges, for example, as expressed by the planetary boundaries?

The purpose of this book, belonging to the series "SpringerBriefs in Biobased Polymers", is to provide an introduction to some of the key challenges of carrying out LCAs of forest products and to suggest some means for handling them. The book can function as a gateway into the literature on LCA of forest products, as it is rich with reference to technical reports and scientific papers. The book is written primarily for LCA practitioners with some previous experience of LCA work, but also less experienced LCA practitioners and others interested in environmental aspects of forests products—such as decision makers confronted with LCA results—can hopefully find it interesting and useful.

As part of the first author's doctoral studies, some of the work presented in this book has previously been published in Sandin (2013, 2015). Figures, tables, and text in these publications might therefore be fully or partly reproduced in the present book.

Gothenburg, Sweden   Gustav Sandin
                                                                     Greg M. Peters
                                                            Magdalena Svanström

References

Sandin G (2013) Improved environmental assessment in the development of wood-based products: capturing impacts of forestry and uncertainties of future product systems. Licentiate thesis, Chalmers University of Technology, 2013:9
Sandin G (2015) Life cycle assessment in the development of forest products: contributions to improved methods and practices. Doctoral thesis, Chalmers University of Technology, serie 3844, ISSN: 0346-718X

Contents

Abbreviations

CH_4	Methane
CO_2	Carbon dioxide
DTA	Demand to availability
EC	European Commission
EU	European Union
EUSES	European Union System for the Evaluation of Substances
FSC	Forest Stewardship Council
GHG	Greenhouse gas
GWP	Global warming potential
ILCD	International reference life cycle data system
ILUC	Indirect land use change
IPCC	Intergovernmental Panel on Climate Change
ISO	International Organization for Standardization
LCA	Life cycle assessment
LCC	Life cycle costing
LCI	Life cycle inventory analysis
LCIA	Life cycle impact assessment
LCSA	Life cycle sustainability assessment
MA	Millennium ecosystem assessment
MCDA	Multi-criteria decision analysis
N_2O	Nitrous dioxide
NIST	American National Institute for Standards and Technology
NPV	Net present value
PEF	Product environmental footprint
PEFC	Programme for the Endorsement of Forest Certification
PVC	Polyvinyl chloride
QRA	Quantitative risk assessment
R&D	Research and development
SETAC	Society of Environmental Toxicology and Chemistry
SIS	Swedish Standards Institute

SLCA	Social life cycle assessment
TEOW	Terrestrial ecoregions of the world
UN	United Nations
UNEP	United Nations Environment Programme
WBCSD	World Business Council for Sustainable Development
WFN	Water Footprint Network
WULCA	Water use in life cycle assessment

Chapter 1
Introduction

Abstract This chapter introduces important environmental and resource crises, describes how increased production and use of forest products are important means for tackling these crises, but stresses that forest products are not necessarily environmentally superior to non-forest products. Furthermore, the chapter underlines the need for carrying out life cycle assessments (LCAs) to evaluate the environmental performance of forests products, and to guide decision making aimed at ensuring that future forest products make sense in environmental terms. The need for improving the methods and practices for LCA of forest products is explained, which leads to the purpose of the book, which is to (i) provide an introduction to some of the key challenges of carrying out LCAs, (ii) suggest means for handling some of the challenges with examples from LCA case studies, and (iii) indicate important research needs to improve LCA methods and practices. It is emphasised that the book can function as a gateway into the literature on LCAs of forest products, and that the main intended audience is the experienced LCA practitioner, but also less experienced LCA practitioners and others interested in the environmental opportunities and challenges of forests products can hopefully find the book interesting and valuable.

Keywords Bio-based economy · Bioeconomy · Circular economy · Wood · Guidance · Planetary boundaries

1.1 The Crises of the Anthropocene

Humankind has entered a new geological epoch, the Anthropocene, in which we are transforming the geology and ecology of the Earth system at a global scale (Steffen et al. 2007). This transformation has been particularly profound following "the great acceleration" after the Second World War—a time period characterised by rapid expansion of the global population, economy, material use and energy use (Steffen et al. 2015a)—with immense consequences for climate (Intergovernmental Panel on Climate Change, IPCC 2013) and ecosystems (Cardinale et al. 2012; Millennium

G. Sandin et al., *Life Cycle Assessment of Forest Products*,
Biobased Polymers, DOI 10.1007/978-3-319-44027-9_1

Ecosystem Assessment, MA 2005; Chapin et al. 2000). The environmental pressures on the Earth system have been summarised in the "planetary boundaries" framework, which suggests nine biophysical boundaries that are intrinsic for the Earth system and which should not be transgressed, so as to avoid risks of abrupt, non-linear, irreversible functional collapses in ecosystems and disastrous consequences for humanity (Steffen et al. 2015b; Rockström et al. 2009). Out of the nine planetary boundaries, at least four are already considered to have been transgressed due to anthropogenic pressures: changes in biosphere integrity, climate change, land-system change and changes in biogeochemical flows (Steffen et al. 2015b). Also others have pointed out the risks of transgressing biophysical thresholds, thereby causing a "state shift in the Earth's biosphere" (Barnosky et al. 2012) or a global "regime shift" in social-ecological systems (Crépin et al. 2012). The global environmental crisis is also shown by "ecological footprint" calculations, which quantify humankind's pressure on the Earth system by accounting for the water and land area needed to meet our demand from nature and assimilate the generated waste. Humankind's ecological footprint is currently estimated to be about 50 % larger than what the Earth can provide for (Global Footprint Network 2015). Another reason for concern is society's dependency on scarce, finite and/or non-renewable resources, for example highlighted in the discussions on "peak oil" (Owen et al. 2010; Sorrell et al. 2009), "peak phosphorus" (Reijnders 2014; Beardsley 2011; Sverdrup and Ragnarsdóttir 2011), "peak rare earth metals" (Ragnarsdóttir et al. 2012) and "peak farmland" (Ausubel et al. 2013).

1.2   New Products Are Needed

The production and use of products are major causes behind the environmental degradation and the dependency on finite resources, and there is widespread international agreement that development of environmentally preferable products is important for addressing these challenges (United Nations, UN 2012). The venture of developing environmentally preferable products is, however, a grand one, as expressed for example by bold targets of reducing the resource intensity per provided service unit (sometimes termed eco-efficiency) in industrial sectors or countries by a factor of 4, 10, 20 or even 50 (Reijnders 1998). The venture is particularly grand if humankind simultaneously intends to reach the recently issued Sustainable Development Goals and increase the standard of living for the world's poor (UN Sustainable Development Knowledge Platform 2016)—which will most probably require increased resource use in the lives of hundreds of millions of people—on a planet expected to be home to almost 10 billion of us by 2050 (UN 2015). Regardless of how much more environmentally efficient the products of tomorrow must be in order for us to stay safely within the planetary boundaries, avoid a state shift in the Earth system, manage finite yet essential resources, support an increasing population and allow development for the less privileged, the message is clear: the environmental impact and resource intensity of products must be considerably reduced.

1.3   The Promise of Forest Products

On the level of overall societal strategies for reducing environmental impact and resource depletion, the idea of industrial ecology asks us to look into how resource flows are organized in nature and to follow the same principles in industrial systems. Such principles have been manifested, for example, in two popular concepts: the circular economy and the bio-based economy (also referred to as "bioeconomy"). The circular economy highlights the need to mimic the circularity of resource flows in natural systems, for example by designing products for recycling. The bio-based economy focuses on building an economy based on biomass, with the underlying notion that the ecosystems that generate the biomass must be sustainably managed in order to maintain long-term productivity and emission assimilation capacity. As an example, the Swedish government has issued a national strategy for the development of a bio-based economy, a transformation described as follows: "the conversion to a bio-based economy entails switching from an economy largely based on fossil resources to a more resource-efficient economy based on renewable resources that are produced through the use of sustainable soil- and water-based ecosystem services" (VINNOVA 2013a, p. 3).

In Sweden, just as in many other countries, a large part of the biomass in question comes from forests. The promise of materials and products derived from forest biomass—henceforth denoted "forest products"—has led to many initiatives for more efficient and more multifaceted use of forest biomass (e.g. in so-called biorefineries) and an increased interest in the research and development (R&D) of new forest products. For example, many European public funding bodies support R&D of new forest products, sometimes as part of wider R&D programmes focussing on biotechnologies and the bioeconomy (see, e.g. BioInnovation 2016; WoodWisdom-net 2016; European Commission, EC 2014a, b, c; VINNOVA 2013b; Formas 2012).

Forest products are, however, not necessarily environmentally preferable compared to non-forest alternatives. For example, forestry and the transformation of non-managed to managed forests can cause biodiversity loss and other environmental degradation, which in turn can undermine ecosystem services that are essential for human livelihood (MA 2005). Also, the subsequent production processes in the forest product value-chain can be demanding both in terms of non-renewable energy and non-forest materials, which can more than offset the benefits of using forest biomass as the main feedstock. For that reason, to create and operate a bio-based economy using forest biomass, there is a need for methods that can assess forest products and handle comparisons with other bio-based products and with fossil-based products.

1.4 Life Cycle Assessment of Forest Products

This book focusses on one of the most widely used tools for the environmental assessment of products: life cycle assessment (LCA), which has been recognised by the industry as the best available method for transparent and reliable assessments of environmental performance (Baitz et al. 2013). The results of LCAs of forest products can be used to guide the transition to a bio-based economy, in terms of improving existing forest products as well as developing new forest products and new forest product value-chains. More specifically, decisions can concern the sourcing of forest biomass, the management of forests, and the development, optimisation and siting of production processes and subsequent processes in the product life cycle (e.g. waste handling). The results of LCAs can also be used for guiding the allocation of public and private funding to future R&D of forest products and for guiding purchases made by consumers or public procurers. Overall, there are many ways in which LCAs can contribute to ensuring that future forest products make sense in environmental terms, and to supporting their market diffusion.

Performing LCAs early in the R&D of forest products is particularly useful as the opportunities for influencing the properties of a product (such as its environmental performance) are greatest in early stages of development and more difficult and expensive later on in the development or once the product has been commercialised (McAloone and Bey 2009; Yang and Shi 2000; Steen 1999; Verganti 1997). As a consequence, to attract public funding for R&D projects aimed at product development, it is sometimes even a requirement to assess the environmental performance of the product under development (Tilche and Galatola 2008).

1.5 Purposes of This Book

This book:

(i) provides an introduction to some of the key challenges of carrying out LCAs of forest products;

(ii) suggests means for handling some of the challenges with examples from LCA case studies of a range of forest products: building materials, fuels, industrial chemicals, textile fibres, etc. (examples are primarily drawn from previous work by the authors);

(iii) indicates important research needs to further improve the methods and practices of LCA of forest products.

The book can function as a gateway into the literature on LCAs of forest products, as it is rich with references to technical reports and scientific papers. The book is intended primarily for the experienced LCA practitioner, but hopefully also less experienced LCA practitioners and others interested in the environmental opportunities and challenges of forests products can find it interesting and useful.

The book differs from more formal guidance documents, such as ISO 14040/14044 (ISO 2006a, b), EN 16760 (SIS 2015), the ILCD handbook (EC 2010) and the PEF guide (EC 2013), in that it focusses on a selection of important challenges of particular relevance for forest products, uses case studies from the scientific literature to explain the challenges and suggest solutions, contains plenty of references to further reading, and points out areas for future research.

1.6 Guide for Readers

Chapter 2 gives an account of the strengths and weaknesses of forest products in terms of environmental performance. Chapter 3 presents LCA methodology. Chapter 4 presents important challenges of carrying out LCAs of forest products, with examples from case studies that illustrate how some of these challenges can be handled. Chapter 5 summarises research needed to further improve the methods and practices of LCAs of forest products. Chapter 6 provides some concluding remarks.

References

Ausubel JH, Wenick IK, Waggoner PA (2013) Peak farmland and the prospect for land sparing. Popul Dev Rev 38(s1):221–242

Baitz M, Albrecht S, Brauner E, Broadbent C, Castellan G, Conrath P et al (2013) LCA's theory and practice: like ebony and ivory living in perfect harmony? Int J Life Cycle Assess 18:5–13

Barnosky AD, Hadly EA, Bascompte J, Berlow EL, Brown JH et al (2012) Approaching a state shift in Earth's biosphere. Nature 486:52–58

Beardsley TM (2011) Peak phosphorus. Bioscience 61(2):91

BioInnovation (2016) BioInnovation—nya biobaserade material, produkter och tjänster. http://www.bioinnovation.se. Accessed Jan 2016

Cardinale BJ, Duffy JE, Gonzalez A, Hooper DU, Perrings C, Venail P et al (2012) Biodiversity loss and its impact on humanity. Nature 486:59–67

Chapin FS III, Zavaleta ES, Eviner VT, Naylor RL, Vitousek PM, Reynolds HL et al (2000) Consequences of changing biodiversity. Nature 405:234–242

Crépin A-S, Biggs R, Polasky S, Troell M, de Zeeuw A (2012) Regime shifts and management. Ecol Econ 84:1522

EC (2010) International reference life cycle data system (ILCD) handbook—general guide for the life cycle assessment—detailed guidance. Joint Research Centre—Institute for Environment and Sustainability. Publications Office of the European Union, Luxembourg

EC (2013) Commission recommendation of 9 April 2013 on the use of common methods to measure and communicate the life cycle environmental performance of products and organisations. http://eur-lex.europa.eu/legal-content/EN/TXT/PDF/?uri=CELEX:32013H0179&from=EN. Accessed Feb 2015

EC (2014a) What is the bioeconomy? http://ec.europa.eu/research/bioeconomy/policy/bioeconomy_en.htm. Accessed Nov 2014

EC (2014b) Biotechnology. http://ec.europa.eu/programmes/horizon2020/en/area/biotechnology. Accessed Nov 2014

EC (2014c). Agriculture & forestry. http://ec.europa.eu/programmes/horizon2020/en/area/agriculture-forestry. Accessed Nov 2014

Formas (2012) Swedish research and innovation strategy for a bio-based economy. Report R3:2012. http://www.formas.se/PageFiles/5074/Strategy_Biobased_Ekonomy_hela.pdf. Accessed Nov 2014

Global Footprint Network (2015) World footprint. http://www.footprintnetwork.org/en/index.php/GFN/page/world_footprint/. Accessed Jan 2016

IPCC (2013) In: Stocker TF, Qin D, Plattner G-K, Tignor M, Allen SK, Boschung J et al (eds) Climate change 2013: the physical science basis. Working group I contribution to the 5th assessment report of the Intergovernmental Panel on Climate Change. Cambridge University Press, Cambridge, UK and New York, NY, USA. http://www.ipcc.ch/report/ar5/wg1/. Accessed Oct 2014

ISO (2006a) 14040: Environmental management—life cycle assessment—requirements and guidelines. International Organisation for Standardisation

ISO (2006b) 14044: Environmental management—life cycle assessment—principles and framework. International Organisation for Standardisation

MA (2005) Ecosystems and human well-being: biodiversity synthesis. World Resources Institute, Washington DC

McAloone TC, Bey N (2009) Environmental improvement through product development: a guide. Danish Environmental Protection Agency, Copenhagen

Owen NA, Inderwildi OR, Kling DA (2010) The status of conventional world oil reserves—hype or cause for concern? Energy Policy 38:4743–4749

Ragnarsdóttir K, Sverdrup HU, Koca D (2012) Assessing long term sustainability of global supply of natural resources and materials. In: Ghenai C (ed) Sustainable development—energy, engineering and technologies—manufacturing and environment. http://www.intechopen.com/books/sustainable-development-energy-engineering-and-technologiesmanufacturing-and-environment/rare-metals-burnoff-rates-versus-system-dynamics-of-metal-sustainability. Accessed Jan 2015

Reijnders L (1998) The factor x debate: setting targets for eco-efficiency. J Ind Ecol 2(13):13–22

Reijnders L (2014) Phosphorus resources, their depletion and conservation, a review. Resour Conserv Recycl 93:32–49

Rockström J, Steffen W, Noone K, Persson Å, Chapin S, Lambin E, et al (2009) Planetary boundaries: exploring the safe operating space for humanity. Ecol Soc 14(2). http://www.ecologyandsociety.org/vol14/iss2/art32/. Accessed Dec 2014

SIS (2015) SS-EN 16760:2015. Bio-based products—life cycle assessment. SIS, Swedish Standards Institute, Stockholm

Sorrell S, Speirs J, Bentley R, Brandt A, Miller R (2009) An assessment of the evidence for a near-term peak in global oil production: a report produced by the Technology and Policy Assessment function of the UK Energy Research Centre. www.ukerc.ac.uk/support/tiki-download_file.php?fileId=283. Accessed Nov 2014

Steen B (1999) A systemic approach to environmental priority strategies in product development (EPS), version 2000—general system characteristics. CPM report 1999:4. http://www.cpm.chalmers.se/document/reports/99/1999_4.pdf. Accessed Jan 2015

Steffen W, Crutzen PJ, McNeill JR (2007) The anthropocene: are humans now overwhelming the great forces of nature? Ambio 36(8):614–621

Steffen W, Broadgate W, Deutsch L, Gaffney O, Cornelia L (2015a) The trajectory of the anthropocene: the great acceleration. Anthropocene Rev 1:1–18

Steffen W, Richardson K, Rockström J, Cornell SE, Fetzer I, Bennett EM et al (2015b) Planetary boundaries: guiding human development on a changing planet. Science. doi:10.1126/science.1259855

Sverdrup HU, Ragnarsdóttir KV (2011) Challenging the planetary boundaries II: assessing the sustainable global population and phosphate supply, using a systems dynamics assessment model. Appl Geochem 26:S307–S310

Tilche A, Galatola M (2008) Life cycle assessment in the European seventh framework programme for research (2007–2013). Int J Life Cycle Assess 13(2):167

UN (2012) Report of the United Nations conference on sustainable development. http://www. uncsd2012.org/content/documents/814UNCSD%20REPORT%20final%20revs.pdf. Accessed Jan 2015

UN (2015) World population prospects, the 2015 revision, the key findings and advance tables. Working paper no. ESA/P/WP.241. United Nations, Department of Economic and Social Affairs, Population Division, New York City, USA

UN Sustainable Development Knowledge Platform (2016) https://sustainabledevelopment.un.org/. Accessed Feb 2016

Verganti R (1997) Leveraging on systemic learning to manage the early phases of product innovation projects. R&D Manage 27(4):377–392

VINNOVA (2013a) A bio-based economy—a strategic research and innovation agenda for new businesses focusing on renewable resources. http://www.vinnova.se/PageFiles/751324632/A% 20BIO-based%20Economy.pdf. Accessed Jan 2016

VINNOVA (2013b) Sectoral R&D programme for the forest-based industry. http://www.vinnova.se/ en/Arkiv/Closed-programmes/Sectoral-RD-Programme-for-the-Forest-based-Industry/. Accessed Mar 2015

WoodWisdom-net (2016) About WoodWisdom-net. http://www.woodwisdom.net. Accessed Jan 2016

Yang Y, Shi L (2000) Integrating environmental impact minimization into conceptual chemical process design—a process systems engineering review. Comput Chem Eng 24:1409–1419

Chapter 2
Strengths and Weaknesses of Forest Products

Abstract This chapter introduces some of the strengths and weaknesses of forest products, for example relating to renewability, biodegradability, climate change, biodiversity loss and water cycle disturbances, indirect land use and land use change. It is explained how the complexities surrounding these topics are key reasons for why environmental assessments are needed to ensure that forest products replacing non-forest products actually reduce environmental impact.

Keywords Transportation · Toxicity · Water use · System boundaries · Carbon balance · ILUC · Life cycle assessment (LCA)

By briefly describing the strengths and weaknesses of forest products, this chapter outlines key drivers for the increased production of, and the development of new, forest products, and key reasons to why environmental assessments are needed to ensure that forest products are environmentally superior.

As described in Chap. 1, the prospect of reduced environmental impact is a common driver for the transformation to a more bio-based society and the development and diffusion of forest products. It is often argued that forest products in general tend to have favourable environmental performance compared to non-forest alternatives (Bergman et al. 2014; Miner et al. 2014; Buyle et al. 2013; Taylor 2013; Werner and Richter 2007), but the use of forest biomass as a feedstock is no guarantee that the end product is environmentally superior to non-forest alternatives.

2.1 Renewability

The potential renewability of forest biomass is an often recognised advantage compared to, for example, those abiotic resources which are subject to scarcity. For forest biomass to be renewable, it must originate from forests which have a constant or growing stock of biomass. Whether this can be claimed depends both on characteristics of the forest (e.g. geographical location) and assumptions in the modelling of the carbon balance of the forest (e.g. the choice of spatial and temporal system boundaries and the baseline chosen to separate the influence of the anthropogenic forestry operations from the natural system). These factors influence also the assessment of climate impact, biodiversity loss and other impact categories. Section 4.3.1 further describes different ways of modelling the carbon balance and how these influence the renewability of forest biomass.

To further demonstrate the complexities involved in determining the renewability of forest biomass, let us elaborate on the situation in Europe. Despite a recent increase in forest biomass stocks in Europe, the stocks may not increase in the future when the products under development today will be produced. The recent increase in boreal biomass stocks is partly a result of long-term recovery from forest degradation in earlier centuries—as noted by Kauppi et al. (2010) for forests in Finland—and this increase may not continue once (if) the historical biomass stocks have been re-established. Indeed, there are signs of a saturation of forest re-growth in Europe (Nabuurs et al. 2013). Moreover, although a higher atmospheric carbon dioxide (CO_2) concentration may induce more biomass growth, disturbances induced by climate change (e.g. increased frequency of forest fires) may eventually result in declining boreal biomass stocks (Kane and Vogel 2009; Kurz et al. 2008). Furthermore, if forest biomass is to replace a substantial share of non-forest (e.g. fossil) resources, the harvesting of forest biomass will have to increase considerably (Narodoslawsky et al. 2008), which may lead to a net decrease also of the forest biomass stocks in Europe. In some regions, such an increased demand may happen considering current energy policies. For example, the European Union (EU) target of achieving a 20 % share of renewable energy in the European energy consumption by 2020 (EU 2007) may cause the demand for European forest biomass to exceed the potential supply (Mantau et al. 2010) and threaten the European forests' capability to function as a carbon sink (Nabuurs et al. 2007). Thus, whether renewability can be claimed depend on the present and future situation in the forest from which the biomass is extracted (as well as methodological choices, as further described in Sect. 4.3.1).

2.2 Biodegradability

Another often recognised benefit of forest biomass is its biodegradability, which means that it will normally not accumulate in nature once it has become a waste material, as some other materials often do, such as many plastics (Derraik 2002). In the disposal stage of forest products, this is sometimes seen as an environmental benefit, although it may not always be a benefit. When forest biomass waste degrades anaerobically, for instance in landfills, part of the carbon is emitted to the atmosphere as methane, a potent greenhouse gas (GHG; Lou and Nair 2009). Globally, methane emissions from landfills may constitute up to 20 % of all anthropogenic methane emissions and 4 % of all anthropogenic GHG emissions

(Frøiland Jensen and Pipatti 2002). The biodegradability may also be problematic in the use phase of forest products and they may therefore require more preservatives, surface treatments and maintenance to meet the same service-life performance as non-forest alternatives. The biodegradability may even make forest biomass unsuitable for some products, such as containers for certain foodstuff.

2.3 Climate Change

Perhaps the most emphasised environmental benefit of forest products concerns their potential role in reducing climate impact. It is commonly claimed that forest products and other bio-based products are carbon neutral, and as a consequence (it is assumed) climate neutral. Such claims may however rely on questionable premises (Agostini et al. 2013; Searchinger et al. 2008). For example, claims of carbon and climate neutrality often presume renewable biomass, which may not always be the case (as discussed in Sect. 2.1). Furthermore, there are mechanisms by which the climate system and forest product systems interact that are not captured by the commonly used methods and practices for climate impact assessment—mechanisms that can contribute both positively and negatively to the climate impact of forest products. The difficulties involved in assessing the climate impact of forest products are further described in Sect. 4.3.2.

2.4 Biodiversity Loss and Water Cycle Disturbances

A potential environmental problem of forest biomass is that it is land- and water-intensive compared with many abiotic resources. Apart from potential problems with renewability and climate impact, as discussed above, poor land and water management can result in a range of other disturbances, including biodiversity loss and water cycle disturbances with subsequent impacts to human health, ecosystem quality and resource availability. Due to looming scarcity of land (Lambin and Meyfroidt 2011) and water (Rockström et al. 2012), such impacts will probably increasingly gain attention, also in countries that seemingly have an abundance of land and water, such as Sweden. Assessment of biodiversity loss and water cycle disturbances are discussed further in Sect. 4.3. It should be noted that the impact category referred to as "water cycle disturbances" in this book, traditionally is referred to as "water use" or "water use impact"; the choice of terminology is further discussed in Sect. 4.3.6.

2.5   Indirect Effects

The environmental impact attributed to a forest product also depends on whether indirect land use and land use change (ILUC) are taken into account. Indirect land use and land use change do not occur at the site of the studied system, but at some other location as a consequence of the activities in the studied system. For example, if land is used for producing a certain product, competition for land increases, which may result in higher commodity prices and therefore more intensive and/or extensive land use and land use change at some other location. Such indirect market-driven effects have been shown to be significant in environmental assessments of biomass feedstocks for biofuels (Berndes et al. 2013; Kløverpris and Mueller 2013; Hertel et al. 2010; Plevin et al. 2010; Searchinger et al. 2008). There has been a greater focus on such indirect effects in studies of bio-based products derived from agricultural feedstocks than in studies of forest products (Ahlgren et al. 2013). However, considering the potentially increasing competition for forest land, there could be significant indirect effects also in future forest product systems. Whether indirect effects should be accounted for depends on the goal and scope of a study, which in turn will affect, for example, whether consequential or attributional assessment approaches are applied (these concepts are further described in Sect. 3.1.1), where consequential approaches more often strive to capture market mechanisms such as indirect land use and land use change. The exclusion of indirect effects may also depend on methodological shortcomings, since the mechanisms behind indirect effects are complex, interlinked, dynamic and uncertain [or, in one word, "wicked" (Coyne 2005)] and thus difficult to quantify (Ahlgren et al. 2013; Berndes et al. 2013).

2.6   Other Aspects of the Environmental Impact of Forest Products

As previously discussed, forest products often require chemical treatment to withstand weathering and degradation, which may lead to exposure of humans and ecosystems to toxic compounds (Werner and Richter 2007). Furthermore, the availability of forest biomass is highly distributed and seasonally variable compared to the availability of many other biotic and abiotic resources, and the energy content of forest biomass is low compared to fossil energy carriers. This can make forest product systems more transport-intensive compared to non-forest product systems, which can considerably influence the environmental performance of forest products (as transportation can be an important contributor to the environmental impact of forest products; Handler et al. 2014). On the other hand, decentralised production can in some situations reduce transportation and the associated environmental impacts. Moreover, the main feedstock of a product is not the only factor determining its environmental impact. For example, in the production and maintenance of forest products, many non-forest materials may be used, sometimes even more (in mass) than used in the production of alternative non-forest products. The amount and type of energy used in the life cycle are also key factors determining a product's environmental impact—factors which can be rather independent of the main feedstock of the product.

2.7 Concluding Remarks

To conclude, the fact that the main feedstock of a product is forest biomass is no guarantee that it is environmentally superior to non-forest alternatives. Many aspects need to be taken into account if one wants to ensure that forest products that replace non-forest alternatives contribute to reduced environmental impact. A number of these aspects are further addressed later on in this book.

References

Agostini A, Giuntilo J, Boulamanti A (2013) Carbon accounting of forest bioenergy: conclusions and recommendations from a critical literature review. JRC Technical Reports, Report EUR 25354 EN. http://iet.jrc.ec.europa.eu/bf-ca/sites/bf-ca/files/files/documents/eur25354en_online-final.pdf. Accessed Dec 2014

Ahlgren S, Björklund A, Ekman A, Karlsson H, Berlin J, Börjesson P (2013) LCA of biorefineries—identification of key issues and methodological recommendations. Report No 2013:25, f3 The Swedish Knowledge Centre for Renewable Transportation Fuels, Sweden. http://www.f3centre.se/sites/default/files/f3_report_2013-25_lca_biorefineries_140710.pdf. Accessed Dec 2014

Bergman R, Puettmann M, Taylor A, Skog KE (2014) The carbon impacts of wood products. Forest Prod J 64(7–8):220–231

Berndes G, Ahlgren S, Börjesson P, Cowie A (2013) Bioenergy and land use change—state of the art. Wiley Interdisc Rev Energy Environ 2:282–303

Buyle M, Braet J, Audenaert A (2013) Life cycle assessment in the construction sector: a review. Renew Sustain Energy Rev 26:379–388

Coyne R (2005) Wicked problems revisited. Des Stud 26:5–17

Derraik JGB (2002) The pollution of the marine environment by plastic debris: a review. Mar Pollut Bull 44:842–852

EU (2007) Brussels European Council 8/9 March 2007: presidency conclusions, 7224/1/07 REV 1. http://register.consilium.europa.eu/doc/srv?l=EN&f=ST%207224%202007%20REV%201. Accessed Jan 2015

Frøiland Jensen JE, Pipatti R (2002) CH$_4$ emissions from solid waste disposal. In: Background papers—IPCC expert meetings on good practice guidance and uncertainty management in national greenhouse gas inventories. Institute for Global Environmental Strategies, Japan

Handler RM, Shonnard DR, Lautala P, Abbas D, Srivastava A (2014) Environmental impacts of roundwood supply chain options in Michigan: life-cycle assessment of harvest and transport stages. J Clean Prod 76:64–73

Hertel T, Alla G, Andrew J, O'Hare M, Plevin R, Kammen D (2010) Global land use and greenhouse gas emissions impacts of U.S. maize ethanol: estimating market-mediated responses. Bioscience 60:223–231

Kane ES, Vogel JG (2009) Patterns of total ecosystem carbon storage with changes in soil temperature in boreal black spruce forests. Ecosystems 12(2):322–335

Kauppi PE, Rautiainen A, Korhonen KT, Lehtonen A (2010) Changing stock of biomass carbon in a boreal forest over 93 years. Forest Ecol Manage 259(7):1239–1244

Kløverpris JH, Mueller S (2013) Baseline time accounting: considering global land use dynamics when estimating the climate impact of indirect land use change caused by biofuels. Int J Life Cycle Assess 18:319–330

Kurz WA, Stintson G, Rampley G (2008) Could increased boreal forest ecosystem productivity offset carbon losses from increased disturbances? Philos Trans R Soc B Biol Sci 363: 2261–2269

Lambin EF, Meyfroidt P (2011) Global land use change, economic globalization, and the looming land scarcity. Proc Natl Acad Sci USA 108(9):3465–3472

Lou XF, Nair J (2009) The impact of landfilling and composting on greenhouse gas emissions—a review. Bioresour Technol 100(16):3792–3798

Mantau U, Saal U, Prins K, Steierer F, Lindner M, Verkerk H et al (2010) Real potential for changes in growth and use of EU forests. EUwood Final report, Hamburg

Miner RA, Abt RC, Bowyer JL, Buford MA, Malmsheimer RW, O'Laughlin J et al (2014) Forest carbon accounting considerations in US bioenergy policy. J Forest 112(6):591–606

Nabuurs G-J, Pussinen A, van Brusselen J, Schelhaas MJ (2007) Future harvesting pressure on European forests. Eur J Forest Res 126:391–400

Nabuurs G-J, Lindner M, Verkerk PJ, Gunia K, Deda P, Michalak R, Grassi G (2013) First signs of carbon sink saturation in European forest biomass. Nature Clim Change 3:792–796

Narodoslawsky M, Niederl-Schmidinger A, Halasz L (2008) Utilising renewable resources economically: new challenges and chances for process development. J Clean Prod 16:164–170

Plevin RJ, O'Hare M, Jones AD, Torn MS, Gibbs HK (2010) The greenhouse gas emissions from market-mediated land use change are uncertain, but potentially much greater than previously estimated. Environ Sci Technol 44:8015–8021

Rockström J, Falkenmark M, Lannerstad M, Karlberg L (2012) The planetary water drama: dual task of feeding humanity and curbing climate change. Geophys Res, Lett 39

Searchinger T, Heimlich R, Houghton RA, Dong F, Elobeid A, Fabiosa J et al (2008) Use of U.S. croplands for biofuels increases greenhouse gases through emission from land-use change. Science 319:1238–1240

Taylor A (2013) Wood is better. Wood Fiber Sci 45(1):1–2

Werner F, Richter K (2007) Wood building products in comparative LCA: a literature review. Int J Life Cycle Assess 12(7):470–479

Chapter 3
LCA Methodology

Abstract This chapter introduces life cycle assessment (LCA) methodology in terms of the four phases of an LCA: goal and scope definition, life cycle inventory analysis, life cycle impact assessment and interpretation. It also introduces the concepts of attributional and consequential LCA, which relate to many of the challenges described in Chap. 4. Finally, it describes the context and limitations of LCA, and its relation to other environmental systems analysis tools, such as life cycle sustainability assessment and quantitative risk assessment.

Keywords Methodology · QRA · Life cycle sustainability assessment (LCSA) · Social life cycle assessment (SLCA) · Life cycle costing (LCC)

LCA is an internationally accepted and widely used method (Baitz et al. 2013; Guinée et al. 2011; Peters 2009) capable of assessing a wide range of environmental impacts over the full life cycle of a product. The LCA procedure consists of four steps, which are usually carried out iteratively to allow for adjustments following new insights (ISO 2006a, b), see Fig. 3.1. Below, each step is described. For further reading about the basics of LCA, we recommend the textbook *The Hitch Hiker's Guide to LCA* (Baumann and Tillman 2004), which is widely us in LCA education at universities.

1. *Goal and scope definition*: The aim of the assessment, the functional unit and the product life cycle are defined, including boundaries to other product systems and the environment. The functional unit is a quantitative unit reflecting the function of the product, which enables the LCA practitioner to compare different products with identical functions. The product life cycle typically includes processes related to resource extraction, manufacturing, use, end-of-life treatment and transportation.
2. *Life cycle inventory analysis (LCI)*: All environmentally relevant material and energy flows between processes within the defined product system, and between the system and the environment and other product systems, are quantified and expressed per functional unit. Flows between the defined system and the environment consist of emissions to different environmental compartments and

© The Author(s) 2016
G. Sandin et al., *Life Cycle Assessment of Forest Products*,
Biobased Polymers, DOI 10.1007/978-3-319-44027-9_3

Fig. 3.1 Schematic
illustration of the four phases
of LCA and their
interconnectedness

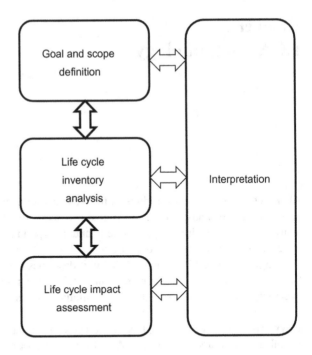

the use of natural resources (including the use of land). These flows are often
termed environmental loads, interventions or stressors.

3. *Life cycle impact assessment (LCIA)*: By means of characterisation methods, the
   LCI data is translated into potential environmental effects, so-called impact
   categories. Traditionally, the focus has been on environmental effects from
   emissions and on global and regional environmental effects, such as climate
   change, stratospheric ozone depletion and eutrophication. Sometimes, LCA
   covers more location-dependent impacts as well, such as eco-toxicity and human
   toxicity. However, there are large uncertainties in the modelling of such impacts
   because they are highly dependent on actual exposure and local or regional
   characteristics (e.g. local flora and fauna, soil structure or presence of other
   substances), which are difficult, or even unfeasible, to account for in LCAs.
   Impact categories can be expressed as inventory-level, midpoint or endpoint
   indicators (see Fig. 3.2). Midpoint indicators (the mandatory and common level
   in LCA) reflect links in the cause-effect chain from activities causing environmental stressors to environmental effects, whereas endpoint indicators (optional
   in LCA) are metrics of actual end effects. For example, the global warming
   potential (GWP) is a midpoint indicator for climate change, as it is based on how
   much an emission influences the radiative forcing. Endpoint indicators for climate change are instead based on how much an emission contributes to possible
   consequences of changed radiative forcing, such as sea level rise, increased
   frequency of extreme weather events or human health consequences of rising

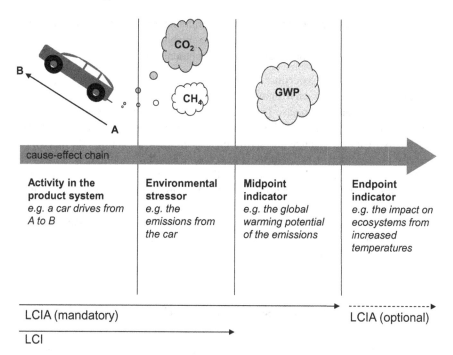

Fig. 3.2 The concept of cause-effect chain and how it relates to LCA methodology, with examples from the cause-effect chain of the climate change impact category

temperatures. Endpoint indicators are sometimes grouped into areas of protection: human health, ecosystem quality, resource availability or (more rarely) man-made environment (Goedkoop et al. 2013; Jolliet et al. 2004).

The LCIA can also include normalisation and weighting. Normalisation can provide understanding of the importance of impacts compared to a reference, by comparing the impact per functional unit to, for example, per capita or aggregated impact in a given area (e.g. global, regional or national) in a certain year (ISO 2006b). Weighting instead compares, and enables the aggregation of, different impact categories on a single yardstick (ISO 2006b). Weighting can be based on, for example, environmental taxes and fees (Finnveden et al. 2006); distances to political goals (Stranddorf et al. 2005); revealed, stated, imputed or political willingness-to-pay for damages (Ahlroth et al. 2011); or end-point models of human, resource and ecosystem damages combined with models of different cultural perspectives (Goedkoop et al. 2013).

4. *Interpretation*: The results of the previous steps are interpreted: the LCIA results are evaluated, taking into account the goal and scope definition (e.g. the system boundaries) and the LCI (e.g. data gaps and data uncertainties), and recommendations are made to the intended audience. The interpretation can include sensitivity and uncertainty analyses (in which the influence of critical or

uncertain system parameters are tested), dominance analysis (in which the contribution of different life cycle processes are analysed), or contribution analysis (in which the contribution of different environmental stressors are analysed).

The above described procedure can be used to assess the impact of a product system on a wide range of environmental concerns. Still, LCA may fail to assess all relevant environmental impacts. The present book describes some of the ongoing research to improve LCA methodology and its practice in various contexts to enable assessments of a wider range of environmental impacts. Nevertheless, it may be necessary to use other assessment tools in certain cases. For example, Sandin et al. (2012) carried out a toxicological evaluation (including a literature study and eco-toxicological testing), in addition to an LCA, to evaluate the toxicological risks of nanoparticles.

To enable understanding of the challenges described in Chap. 3, one aspect of LCA methodology needs to be elaborated on in more detail: the choice between attributional and consequential modelling.

3.1   Attributional or Consequential LCA?

The consequential-attributional controversy is a topic of discussion in the LCA research community (Plevin et al. 2014; Suh and Yang 2014; Earles and Halog 2011; Ekvall and Weidema 2004; Tillman 2000), possibly exacerbated by inconsistent recommendations in guiding documents such as the ILCD handbook (Ekvall et al. 2016). Traditionally, LCA has relied on attributional (also called descriptive or accounting) approaches, which (most often) means that the LCA considers the immediate physical flows (emissions and resource use) occurring at the location of the life cycle processes. Attributional approaches typically imply that the LCA maps the average impact of the studied product system per delivered functional unit. A consequential (also called change-oriented) approach, on the other hand, seeks to map the change of physical flows occurring as a consequence of a decision (Zamagni et al. 2012; Earles and Halog 2011; Ekvall and Weidema 2004). This can also be described as the consequences of a change in production output, i.e. what the environmental consequence would be if more or less functional units were provided. A consequential approach entails inclusion of effects not necessarily physically connected to the product system, but occurring due to, for example, market mechanisms (Earles and Halog 2011; Ekvall and Weidema 2004). Section 2.5 described one such market mechanism: indirect land use and land use change. Another way of describing the distinction between attributional and consequential approaches is that attributional approaches assumes the surrounding world is static, i.e. other technical systems are not influenced by the studied product system, whereas consequential approaches assumes the surrounding world is dynamic, i.e. there are indirect (secondary) effects occurring in other technical systems as a consequence of the studied product system. The choice between an attributional and a consequential approach determines, for example, which processes to include within the system boundaries, which LCI data to use (see the next paragraph) and how to handle multi-functional processes (see Sect. 4.2). Later in this book, there are several examples of when consequential and attributional approaches lead to different LCIA results for forest products. Indeed, it has been argued that the choice between consequential and attributional approaches is particularly important for bio-based products (Pawelzik et al. 2013). See Zamagni et al. (2012) for a further review of consequential and attributional LCA methodology.

One important concern related to the choice between attributional and consequential approaches is whether to use average or marginal LCI data. For example, when the studied product requires electricity for its production, it is common to use average LCI data, i.e. data on the annual average emission per unit of electricity produced in the country or region of the production site. However, marginal LCI data can also be used, which are emission data on the marginal source for electricity, i.e. the technology that is expected to respond to a change in demand. The marginal technology is most often considered to be the utilised technology with the highest operating cost (also called marginal cost) or the unutilised technology with the lowest operating cost (Lund et al. 2010). However, some authors have proposed that in markets constrained by regulation, the planned or predicted technology should rather be considered the marginal one (Schmidt et al. 2011). Typically, average data are used for attributional studies, and marginal data for consequential studies (Ekvall and Weidema 2004). The use of marginal data is based on the consequential logic that if the product is not produced, the marginal technology will not be utilised. In many countries, the marginal technology for electricity generation is coal power, which only contributes to the electricity mix when demand is particularly high. As emissions from coal power can be much higher than emissions from the average electricity generation (which may be dominated by, e.g. hydro or nuclear power), the choice between average and marginal LCI data can significantly influence LCIA results. It can, however, be difficult to determine the marginal technology (Mathiesen et al. 2009). For example, the short-term marginal technology (e.g. at a particular time of the day, or a particular time of the year) may be different from the long-term marginal technology (e.g. annually). Thus, the choice between, and the selection of, average or marginal LCI data is a much discussed aspect of LCA methodology.

3.2 Context and Limitations of LCA

The focus of this book is on one particular environmental assessment tool, LCA. For many contexts it can be more appropriate to use other assessment tools, sometimes to complement the use of LCA. The need for using several complementary tools in environmental assessments has been emphasised before; for example, by Chico et al. (2013) for assessing water cycle disturbances and by Buonocore et al. (2014) for environmental assessments of forestry operations. It is thus important to stress that, just as the use of LCA must be adapted to the specific context of the study—for example in terms of the choice of method for handling multi-functionality (Sect. 4.2), of impact assessment methods (Sect. 4.3) and of scenario modelling approach (Sect. 4.1)—the choice between LCA and other tools must also be made in light of the specific context. In particular, alternative environmental assessment or risk assessment tools may be needed until certain dimensions of the LCA method have been further developed, such as sufficiently location-specific impact assessment. One such example is given in Sandin et al. (2012), in which we needed to—because of inadequate LCA methods—use complementary toxicological testing to assess the environmental risks of nanoparticles.

3.2.1   Relationship Between LCA and Other Assessment Tools

LCA has many relatives in the family of environmental systems analysis tools. Some of the key ones are social life cycle assessment (SLCA) and life cycle costing (LCC). SLCA and LCC are sometimes grouped together with LCA under the heading of life cycle sustainability assessment (LCSA), since sustainability traditionally covers impacts on human society, the economy and nature, respectively (Kloepffer 2009).

LCC is perhaps more closely related to the process of net present value (NPV) analysis in that the focus is on financial impacts, and discounting of future cash flows is commonplace. Discounting the future is not a popular activity in LCA although some models of impacts on humans do apply age-based discounting of effects on human health (Kobayashi et al. 2015). LCC attempts to be more comprehensive than some NPV calculations by taking a life cycle perspective of products and services. Key guidance documents for LCC are provided by the American National Institute for Standards and Technology (NIST) (Fuller and Petersen 1996) and the Society of Environmental Toxicology and Chemistry (SETAC) (Swarr et al. 2011). Naturally, if the person performing LCC does not include externalities in the calculations, then LCC will only reflect environmental costs which have been monetised by legislation (such as carbon trading schemes). In this sense, LCA and LCC are very different and typically very useful for covering each others' blind spots.

SLCA is a tool that is currently relatively immature compared to LCA and LCC, and therefore it is the subject of much methodological dynamism. In brief, it follows the same four-step structure as LCA but introduces different indicators. The UN Environment Programme (UNEP)/SETAC SLCA guidelines (Benoît and Mazijn 2009) propose a list of indicators based on the kinds of stakeholders affected by product systems (workers, local communities, society, consumers and value-chain actors) and the kinds of social impacts that seem relevant (child labour, unfair salaries, unsafe working conditions, corruption, etc.). Given the length of this list, the analyst still has an important task of identifying which indicators to use, and this is the subject of much critical debate today (Arvidsson et al. 2015; Sandin et al.

2011). Nevertheless, the opportunity to assess social sustainability impacts along product life cycles is keenly sought by many stakeholders.

LCA, LCC, SLCA and LCSA are all attempts to try and lift the thinking of decision makers from a limited focus on financial costs (e.g. NPV) when developing products, processes and policies, to allow the decision makers to adopt a better informed and broader worldview. These life cycle tools incorporate a larger conception of the technical system that delivers the products and services we ask for, and a broader conception of the environment in which the products and services are created. As a consequence, they are better (than assessment tools with a more limited perspective) at avoiding problem-shifting and at recognizing trade-offs between different dimensions of sustainability. A recent review by Beaulieu et al. (2015) suggested that these tools and others can help lead the transition of our economy to a more sustainable basis, and provides a lengthy discussion of the relationship between these and other tools, and related sustainability concepts. Finnveden and Moberg (2005) also provide a useful review of environmental system analysis tools.

Another tool that has a long and multifaceted relationship with LCA is quantitative risk assessment (QRA). The first attempts to include chemical hazards in an LCA framework were built on risk-based air and water quality guideline values. The characterisation process was basically a division of an emitted mass of a substance by a guideline value (e.g. drinking water quality), creating a result denominated in volumetric units. The "dilution volume" was then proportional to the potential environmental impact associated with the discharge. This method is preserved in current usage by the Water Footprint Network (WFN), in the form of the "grey water footprint". A more advanced symbiosis has developed between LCIA and level 3 fugacity models developed by Donald Mackay and others (Mackay 2001). The most relevant one of these for contemporary LCA analysts is Usetox 2.0, which consists of a number of nested compartments at local, continental and global scales. This is a descendent of models like the European Union System for the Evaluation of Substances (EUSES) which were originally developed for the regulation of chemicals and other policy-making applications of QRA rather than for LCA (Vermeire et al. 2005). LCA practitioners frequently dip into the methods developed for QRA to enhance the accuracy of LCIA when local exposures play an important role in the overall impacts of a product or service (see, e.g. Harder et al. 2015; Heimersson et al. 2014; Andersson et al. 2014). The main differences between normal QRA for chemical hazards and the models used to generate LCIA characterisation factors are that the LCIA models are typically designed to reflect average emissions (rather than catastrophic conditions), chronic exposure (rather than acute), risks to the general population (rather than occupational risks) and a marginal increase in exposure (rather than taking background concentrations into account). Relatedly, the results of QRA models are frequently interpreted by comparison with an acceptable limit, whereas LCA is frequently concerned with comparing one product with another. So while QRA usually sets out to answer different questions to LCA, the two perspectives are complementary, as witnessed by the many attempts to hybridise them in some way (Harder et al. 2015).

References

Ahlroth S, Nilsson M, Finnveden G, Hjelm O, Hoschchorner E (2011) Weighting and valuation in selected environmental systems analysis tools—suggestions for further developments. J Clean Prod 19(2–3):145–156

Andersson H, Harder R, Peters G, Cousins I (2014) SUPFES: environmental risk assessment on short-chain per- and polyfluoroalkyl substances applied to land in municipal sewage sludge. 6th International Workshop on Per- and Polyfluorinated Alkyl Substances – PFAs, Idstein, Germany, 15–18 June

Arvidsson R, Baumann H, Hildenbrand J (2015) On the scientific justification of the use of working hours, child labour and property rights in social life cycle assessment: three topical reviews. Int J Life Cycle Assess 20(2):161–173

Baitz M, Albrecht S, Brauner E, Broadbent C, Castellan G, Conrath P et al (2013) LCA's theory and practice: like ebony and ivory living in perfect harmony? Int J Life Cycle Assess 18:5–13

Baumann H, Tillman A-M (2004) The hitch hiker's guide to LCA. Studentlitteratur, Lund

Beaulieu L, van Durme G, Arpin M-L, Reveret J-P, Margni M, Fallaha S (2015) Circular economy: a critical review of concepts. CIRAIG, Montreal. ISBN 978-2-9815420-0-7

Benoît C, Mazijn B (eds) (2009) Guidelines for social life cycle assessment of products. http://www.unep.org/publications/search/pub_details_s.asp?ID=4102. Accessed Feb 2015

Buonocore E, Häyhä T, Paletto A, Franzese PP (2014) Assessing environmental costs and impacts of forestry activities: a multi-method approach to environmental accounting. Ecol Model 271:10–20

Chico D, Aldaya MM, Garrido A (2013) A water footprint assessment of a pair of jeans: the influence of agricultural policies on the sustainability of consumer products. J Clean Prod 57:238–248

Earles JM, Halog A (2011) Consequential life cycle assessment: a review. Int J Life Cycle Assess 16:445–453

Ekvall T, Weidema BP (2004) System boundaries and input data in consequential life cycle inventory analysis. Int J Life Cycle Assess 9(3):161–171

Ekvall T, Azapagic A, Finnveden G, Rydberg T, Weidema BP, Zamagni A (2016) Attributional and consequential LCA in the ILCD handbook. Int J Life Cycle Assess 20:1–4

Finnveden G, Moberg Å (2005) Environmental systems analysis tools—an overview. J Clean Prod 13(12):1165–1173

Finnveden G, Eldh P, Johansson J (2006) Weighting in LCA based on ecotaxes: development of a mid-point method and experiences from case studies. Int J Life Cycle Assess 11:81–88

Fuller S, Petersen S (1996) Life-cycle costing manual. Handbook 135. National Institute for Standards and Technology, Washington

Goedkoop M, Heijungs R, Huijbregts M, de Schryver A, Struijs J, van Zelm R (2013) ReCiPe 2008 (version 1.08)—report I: characterisation (updated May 2013). http://www.lcia-recipe.net. Accessed Nov 2014

Guinée JB, Heijungs R, Huppes G, Zamagni A, Masoni P, Buonamici R et al (2011) Life cycle assessment: past, present, and future. Environ Sci Technol 45:90–96

Harder R, Andersson H, Molander S, Svanström M, Peters G (2015) Review of environmental assessment case studies featuring elements of risk assessment and life cycle assessment. Environ Sci Technol 49(22):13083–13093

Heimersson S, Harder R, Peters GM, Svanström M (2014) Including pathogen risk in life cycle assessment of wastewater management. Part 2: Quantitative comparison of pathogen risk to other impacts on human health. Environ Sci Technol 48:9446–9453

ISO (2006a) 14040: Environmental management—life cycle assessment—requirements and guidelines. International Organisation for Standardisation

ISO (2006b) 14044: Environmental management—life cycle assessment—principles and framework. International Organisation for Standardisation

Jolliet O, Müller-Wenk R, Bare J, Brent A, Goedkoop M, Heijungs R et al (2004) The LCIA midpoint-damage framework of the UNEP-SETAC life cycle initiative. Int J Life Cycle Assess 9:394–404

Kloepffer W (2009) Life cycle sustainability assessment of products. Int J Life Cycle Assess 13 (2):89–95

Kobayashi Y, Peters G, Ashbolt N, Khan S (2015) Assessing burden of disease as disability adjusted life years in life cycle assessment. Sci Total Environ 530–531:120–128

Lund H, Mathiesen BV, Christensen P, Schmidt JH (2010) Energy system analysis of marginal electricity supply in consequential LCA. Int J Life Cycle Assess 15:260–271

Mackay D (2001) Multimedia environmental models. Lewis Publishers. ISBN: 1-56670-542-8

Mathiesen BV, Münster M, Fruergaard T (2009) Uncertainties related to the identification of the marginal technology in consequential life cycle assessments. J Clean Prod 17:1331–1338

Pawelzik P, Carus M, Hotchkiss J, Narayan R, Selke S, Wellisch M et al (2013) Critical aspects in the life cycle assessment (LCA) of bio-based materials—reviewing methodologies and deriving recommendations. Resour Conserv Recycl 73:211–228

Peters GM (2009) Popularize or publish? Growth in Australia. Int J Life Cycle Assess 14:503–507

Plevin RJ, Delucchi MA, Creutzig F (2014) Response to "On the uncanny capabilities of consequential LCA" by Sangwon Suh and Yi Yang (Int J Life Cycle Assess, doi:10.1007/s11367-014-0739-9). Int J Life Cycle Assess 19, 1559–1560

Sandin G, Peters GM, Pilgård A, Svanström M, Westin M (2011) Integrating sustainability consideration into product development: a practical tool for identifying critical social sustainability indicators and experiences from real case application. In: Finkbeiner M (ed) Towards life cycle sustainability management, Springer, Dordrecht, pp 3–14

Sandin G, Pilgård A, Peters GM, Svanström M, Ahniyaz A, Fornara A, Johansson Salazar-Sandoval E, Xu Y (2012) Environmental evaluation of a clear coating for wood: toxicological testing and life cycle assessment. In: Conference proceedings of the 8th International PRA Woodcoatings Congress, Amsterdam, the Netherlands

Schmidt JH, Merciai S, Thrane M, Dalgaard R (2011) Inventory of country specific electricity in LCA—consequential and attributional scenarios, Methodology report. http://lca-net.com/files/Inventory_of_country_specific_electricity_in_LCA_Methodology_report_20110909.pdf. Accessed Jan 2015

Stranddorf HK, Hoffmann L, Schmidt A (2005) Update on impact categories, normalisation and weighting in LCA–selected EDIP97 data. Environmental Project Nr. 995 2005, Danish Environmental Protection Agency, Copenhagen. http://www2.mst.dk/udgiv/publications/2005/87-7614-570-0/pdf/87-7614-571-9.pdf. Accessed Oct 2014

Suh S, Yang Y (2014) On the uncanny capabilities of consequential LCA. Int J Life Cycle Assess 19:1179–1184

Swarr TE, Hunkeler D, Klöpffer W, Pesonen H-L, Ciroth A, Brent AC, Pagan R (2011) Environmental life cycle costing: a code of practice. Society of Environmental Toxicology and Chemistry (SETAC), Pensacola

Tillman AM (2000) Significance of decision-making for LCA methodology. Environ Impact Assess Rev 20(1):113–123

Vermeire T, Rikken M, Attias L, Boccardi P, Boeije G, Brooke D et al (2005) European Union system for the evaluation of substances: the second version. Chemosphere 59:473–485

Zamagni A, Guinée J, Heijungs R, Masoni P, Raggi A (2012) Lights and shadows in consequential LCA. Int J Life Cycle Assess 17:904–918

Chapter 4
LCA of Forest Products—Challenges and Solutions

Abstract This chapter provides an extensive walkthrough of the important challenges encountered when carrying out life cycle assessment (LCA) of forest products, and proposes some solutions to these challenges, with examples from the scientific literature and technical reports. The topics include: modelling future and/or uncertain product systems, handling multi-functionality (i.e., allocation problems), inventory analysis and impact assessment (carbon flow modelling, assessing climate impact, biodiversity loss, water cycle disturbances and energy use), managing trade-offs and connecting the LCA work to global environmental challenges, and integrating LCA work in the R&D of new products.

Keywords Inventory data · Consequential · Attributional · Temporal modelling · Spatial modelling · Baseline · Characterisation method · Carbon footprint · Climate change · Land use · Water use · Planetary boundaries

In LCAs of forest products, some typical challenges appear. For instance, some of these relate to the forest feedstock and its origin, some relate to the modelling of new, emergent, multi-functional and/or long-lived product systems, some relate to the interpretation of results, and some relate to the integration of LCA work in various contexts. In the following, some such challenges are described, together with discussions of potential solutions. First, the chapter addresses challenges primarily related to the first LCA phase, the goal and scope definition: the modelling of future and/or uncertain product systems and the handling of multi-functionality. Then, the chapter addresses challenges primarily related to the second and third LCA phases, the inventory analysis and the impact assessment: the modelling of carbon flows in the forest, and the assessment of climate impact, biodiversity loss, water cycle disturbances and energy use. Thirdly, the chapter addresses challenges primarily related to the fourth LCA phase, the interpretation: the management of trade-offs between impact categories and how to relate LCIA results to global environmental challenges. Finally, the chapter addresses the challenge of integrating LCA work in the R&D of new products, with a particular focus on inter-organisational R&D projects. Not all these challenges are exclusive for forest products, but they are common challenges for LCA practitioners involved in studies of forest products.

© The Author(s) 2016
G. Sandin et al., *Life Cycle Assessment of Forest Products*, Biobased Polymers, DOI 10.1007/978-3-319-44027-9_4

4.1 Modelling Future and/or Uncertain Product Systems

In the envisioned future bio-based society much will be different from today. Many already established forest products will be further developed, and factors influencing their environmental performance will change—including background systems (the production of electricity, fuels and input materials) and land-use activities (due to changes in the competition for land). Uncertainties in forest product systems and in their environmental performance are even more pronounced for the products under development today—products yet to enter the market. Further, methodological choices also influence the outcome of the LCA and thus also pose a type of uncertainty. It is important to capture all of these uncertainties in LCAs of forest products if the aim is to develop product systems that contribute to reduced environmental impact regardless of future world development and whose perceived environmental benefits are not dependent on arbitrary methodological choices. To capture these types of uncertainties in LCAs, scenario modelling and sensitivity analysis are commonly used.

4.1.1 Definitions and Classifications

A scenario in LCA has been defined as "a description of a possible future situation relevant for specific LCA applications, based on specific assumptions about the future, and (when relevant) also including the presentation of the development from the present to the future" (Pesonen et al. 2000, p. 23). There are various systems for classifying scenarios in LCA. Börjeson et al. (2005, 2006) distinguished between predictive (what will happen?), explorative (what can happen?) and normative (how can a specific target be reached?) scenarios. For each category, Börjeson and colleagues made a further breakdown: predictive scenarios can be forecasts or what-if scenarios, explorative scenarios can be external or strategic, and normative scenarios can be preserving or transforming. Pesonen et al. (2000) distinguished between what-if and cornerstone scenarios, where what-if scenarios are used to compare the environmental consequences of choosing between well-defined options in a well-known and simple situation, while cornerstone scenarios are used to compare options in a more unknown and complex situation to increase the understanding of the studied system (e.g. in the context of product development, as recognised by Pesonen and colleagues). Scenario analysis has been pointed out to be particularly useful in LCAs dealing with strategic decisions (Guinée et al. 2002) with non-marginal implications for technological systems (Heijungs et al. 2009). Non-marginal implications in this case refer to situations in which "technologies can no longer be characterized with constant coefficients; capacities of technologies is [sic] no longer constant; the change in economic structure will affect prices and preferences, and hence induce change in life styles; background concentrations of pollutants will change" (Heijungs et al. 2009, p. 61). Furthermore, Heijungs and colleagues emphasised that scenarios of such changes can be set up endogenously, as the outcome of models, or exogenously, as the outcome of "creative or explorative thinking" (p. 62), which resembles the explorative scenario classification by Börjeson et al. (2005). See Börjeson et al. (2006) for a more extensive review of scenario classification.

4.1.2 Examples from the Literature

The literature provides a rich plethora of examples for how to model scenarios or carry out sensitivity analysis in LCAs in order to capture uncertainties of different kinds, for example uncertainties related to the surrounding world, the scope of the study, technological assumptions, LCI data, methods for handling multi-functionality or impact assessment methods (see, e.g. Grant et al. 2014; Peters et al. 2013; Bhattacharyya et al. 2013; Cellura et al. 2011; Cherubini et al. 2011; Mathiesen et al. 2009; Ardente et al. 2008; Spielmann et al. 2005). Some of these papers refer to the different setups tested in a sensitivity analysis as "scenarios", thus scenario modelling and sensitivity analysis are not necessarily distinctly different concepts. Others have also acknowledged that scenario analysis can encompass sensitivity testing (Börjeson et al. 2006). In the below bullet list, we introduce some scenario generation techniques suitable for LCAs and give some examples of case studies. For further details, the reader is referred to the original paper/report.

- Börjeson et al. (2006) categorised techniques for creating scenarios into generating, integrating and consistency techniques. Among generating techniques, they listed surveys, workshops and Delphi methods (a structured process to capture and harmonise the opinions of experts on an issue) as means for collecting and reviewing ideas, opinions, input data, assumptions, model calculations and model results. Integrating techniques concern how to integrate the parts (e.g. collected by generating techniques) into wholes (i.e. final scenarios). Among consistency techniques—means for ensuring consistency between or within scenarios—they provided two examples: cross-impact analysis for checking consistency in terms of the causality of the parameters defining the scenario, and morphological field analysis for checking consistency in terms of the possibility that the parameters defining a scenario can co-exist.
- Spielmann et al. (2005) proposed a method for generating a set of "possible, consistent and diverse cornerstone scenarios representing future developments of an entire LCI product system" (p. 326). The method outlines a series of steps for (i) selecting socio-economic and technological factors that can be expected to influence each process in the studied product system, (ii) analysing how each process can be influenced in terms of the LCI data, and (iii) integrating the influence on each unit process into cornerstone scenarios for the entire product system.
- Mathiesen et al. (2009) proposed a method for scenario modelling in consequential studies of energy systems, in which different marginal technologies are assumed in the generation of a set of fundamentally different future scenarios.

- Grant et al. (2014) generated a set of methodologically differentiated scenarios by using five different service life models to account for the uncertainties of predicting the future service life in LCAs of building envelopes.
- Peters et al. (2013) used triangulation to handle scope uncertainty in an LCA of air to water machines, by comparing measured, claimed and theoretical estimates of the key variable influencing the performance of the studied product.
- Bhattacharyya et al. (2013) used sensitivity analysis to study the future service life's influence on the climate impact of forest products.
- Cellura et al. (2011) used sensitivity analysis to test the influence of LCI data, transportation assumptions and characterisation methods in an LCA of roof tiles.
- Sandin et al. (2014b) generated end-of-life scenarios in the assessment of construction products by making different assumptions about the nature of future technologies for transportation, demolition of constructions and waste handling. Additionally, the scenarios tested the influence of using either attributional or consequential modelling approaches. The scenarios enabled a study of how the product system under development (in this case a glue-laminated wooden beam) would perform compared to an alternative product system (in this case a steel frame) in (i) a future in which technologies have roughly the same environmental impact as today's technologies, and (ii) a future in which technologies have considerably lower environmental impact than today's technologies (i.e. an example of cornerstone scenarios). If the developed product performs better in all scenarios, it is probably a long-term environmentally and commercially attractive alternative in many types of decision contexts. On the other hand, if there turns out to be small or non-existent environmental benefits of the developed product in a future dominated by low- or high-impact technologies, or in a study based on a certain modelling approach, further development of the product (and/or careful planning of the supply chain design) is probably appropriate both for environmental and commercial reasons.
- Ardente et al. (2008) used sensitivity analysis to test the influence of different methods for handling multi-functionality problems (multi-functionality problems are further discussed in Sect. 4.2) and different transportation assumptions in an LCA comparison of insulation materials. Similarly, Cherubini et al. (2011) tested the influence of the method for handling multi-functionality problems in LCAs of biorefinery products.
- Similarly as Ardente et al. (2008) and Cherubini et al. (2011), Sandin et al. (2015c) used scenarios to test the implications of the method for handling multi-functionality in LCAs of biorefinery products. This was done in relation to four decision contexts relevant for forest products produced in biorefineries. The decision contexts were generated by varying two contextual dimensions deemed to be defining in terms of LCA methodology: contexts relating to *specific* or *general* product systems, and contexts relating to *minor* or *major* modifications of product systems. Minor modifications can for instance be incremental changes of an existing product system or changes induced by choices made by individual consumers. Major modifications can for instance be changes caused by the development of a new product system or by policy-making. Since the range of

the outcomes of using different methods for handling multi-functionality can—as long as the use of each method is reasonably justified—be seen as a methodological uncertainty of the assessment, the paper serves as an example of how a structured generation of a set of decision contexts can be used to evaluate the possible influence of methodological uncertainty in LCAs.

- Røyne et al. (2016) explored uncertainties imposed by possible climate impact assessment practices in LCAs of forest products (Sect. 4.3 further describes the challenges of climate impact assessment of forest products). The uncertainties were explored by varying two parameters: the type of product and the impact assessment practice. Product type was tested by studying two forest products with very different life lengths: an automotive fuel and a building. A range of possible climate impact assessment practices were tested, and then, the paper showed the outcome of the common practice at the time and the practices yielding the lowest and highest LCIA results, respectively (which can be classified as cornerstone scenarios). A set of distinctly different decision contexts were generated in order to explore potential implications of the choice of climate impact assessment practice for decision-making. As in Sandin et al. (2015c), this was done by varying two contextual dimensions that are considered important for methodological choices (Finnveden et al. 2009; Tillman 2000): whether the LCA is performed on a specific or a general product, and whether or not the LCA compares products. By showing the span of possible LCIA results if LCA practitioners were to use other impact assessment practices than commonly used, the paper serves as an example of how uncertainties arising from the choice of impact assessment method can be tested in LCAs.
- Sandin et al. (2013) set up scenarios to map uncertainties of the location of a future forest product system by varying two factors: the market demand for the product (affecting the scale of the product system) and the expected competition for land (affecting the need to turn to previously unused forest land). The outcome of this type of scenario modelling can for example show to what extent the sourcing of the main feedstock of the forest product—where it comes from and how the extraction has been managed—is crucial for the product to be an environmentally preferable alternative. This information can, for example, help to design the supply chain.
- Arvidsson et al. (2015b) assessed the life cycle environmental impacts of cellulose nanofibrils produced via different routes. In order to investigate the inherent uncertainty in the emergent products systems, a thorough sensitivity analysis was made that considered variations in pulp production, process electricity use, transport modes and distances, ethanol production, electricity mix, heat source, and the introduction of solvent heat and material recovery. Further, the potential influence of scale-up and changes in background systems were discussed.

Previous research has recognised some inherent uncertainties of future product systems that prospective LCAs should consider, for example uncertainties in the LCI data for commercial scale processes that today only exist at laboratory scale (Hetherington et al. 2014). The research presented in the present book highlights some additional uncertainties that appear to be typical for future product systems in general and future *forest* product systems in particular, and thus are important to consider in LCAs of such systems, and important to capture in scenario modelling and sensitivity analysis. These include: the type of technology assumed in the disposal of the studied product and the technology assumed for the substituted processes when substitution is used for handling multi-functionality (see Sects. 4.1.3 and 4.2), the location of processes (in particular forestry operations) and the occurrence of land use change (see Sect. 4.3). Some of these uncertainties are present also in existing product systems. In the following subsection, a particular modelling uncertainty is discussed in greater detail: the uncertainty associated with the end-of-life modelling of long-lived products.

It should be noted that the appropriate approach for scenario modelling and sensitivity analysis will always depend on the unique context of each study: the approach must generate some understanding that supports the LCA practitioner in reaching the goal of the study, ultimately improving decision-making based on the study. It is thus important that any scenario modelling approach is selected for, or adapted to, the specific context of each study.

4.1.3 End-of-Life Modelling of Long-lived Products

Buildings and other constructions are an important use of forest biomass. The end-of-life processes (e.g. demolition and disposal) of such products manufactured today will often take place in 50–100 years (Frijia et al. 2011). The nature of such processes is highly uncertain because of technological change (Du et al. 2014; Frischknecht et al. 2009a). This time-dependent technological uncertainty has been acknowledged as being a common challenge for LCAs in the construction industry (Singh et al. 2011; Verbeeck and Hens 2007). Still, it is an often neglected uncertainty in LCAs of construction products. The end-of-life technologies of today are often assumed without any explicit explanation, also when the aim is to support decisions concerning contemporary or future constructions with end-of-life processes occurring in a distant future (e.g. Habert et al. 2012; Bribián et al. 2011; Persson et al. 2006; Lundie et al. 2004). There are exceptions, for example Bouhaya et al. (2009) used scenarios to account for different possible end-of-life technologies in an LCA of a bridge, Du et al. (2014) used sensitivity analysis to test different steel recycling rates for the end-of-life handling in an LCA of five bridge designs, and Garcia and Freire (2014) tested the influence of assuming either incineration or landfill as the end-of-life option in carbon footprint calculations of wood-based panels.

Accounting for end-of-life uncertainties is particularly important when end-of-life processes can be expected to strongly influence a product's environmental impact. Efficient recycling in the disposal of buildings may save energy corresponding to 29 % of the energy use in manufacturing and transportation of the construction materials (Blengini 2009) or 15 % of the total energy use of a building's life cycle (Thormark 2002). Although the use phase has been said to contribute 60–90 % of a building's environmental impact (Buyle et al. 2013; Cuéllar-Franca and Azapagic 2012; Ortiz et al. 2010), the relative contribution from end-of-life processes is growing because of increasingly energy-efficient buildings (Dixit et al. 2012). Furthermore, it has been argued that inadequately defined functional units have led to overstated energy usage in the use phase (Frijia et al. 2011), which implies that the relative contribution from end-of-life processes has been understated. The sheer amount of construction materials existing in society has also been used as an argument for why the end-of-life handling of such materials is environmentally important (Bribián et al. 2011; Singh et al. 2011; Blengini 2009).

So there are many reasons for improving the modelling of end-of-life processes in LCAs of long-lived forest products. Sandin et al. (2014b) provide an example of how the uncertainties of end-of-life processes can be captured in LCA by scenario analysis. In doing this, Sandin and colleagues revealed four factors of the end-of-life modelling that are especially critical for the LCA results: whether end-of-life phases are considered at all, whether recycling or incineration is assumed in the disposal, whether consequential or attributional end-of-life mod- elling approaches are used, and whether today's average technology or a low-impact technology is assumed for the substituted technology. The factors are therefore particularly important to handle carefully when modelling end-of-life processes of long-lived forest products. Furthermore, Sandin and colleagues showed that when consequential end-of-life modelling with substitution is used, it appears to be particularly important to set up several distinctly different scenarios, because the outcome of the assessment largely depends on speculative assumptions regarding the type of substituted technology, as has been recognised before (Pelletier et al. 2015; Zamagni et al. 2012; Mathiesen et al. 2009; Heijungs and Guinée 2007). Until there is a clearer consensus in the LCA community on when to use attributional and consequential approaches, there is a general need to use both approaches simultaneously—as long as each approach is reasonably justified in the context of the study—to facilitate robust LCA-based decision making. There is also a need to generate several distinctly different scenarios for each approach, for example regarding the type of substituted technology for consequential end-of-life modelling. Use of both approaches, and several scenarios in each approach, within a single case study, can improve our understanding of the approaches and under what circumstances the selection of approach matters most.

4.2 Handling Multi-functionality

Another typical feature of forest product systems is multi-functional processes (also called multi-output processes). For example: forestry often provides timber, pulp- wood and fuelwood; the subsequent production often yields several products (e.g. in the case of biorefineries: paper pulp, fuels, heat and chemicals); and the waste handling may provide recyclable materials, reusable materials, heat and/or elec- tricity. As biorefineries become more common and more integrated, and more

Fig. 4.1 An example of a
multi-functionality problem,
i.e. how to allocate emission
X from the factory between
products A, B and C

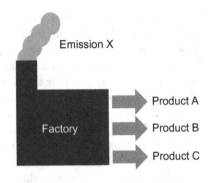

products are produced in each biorefinery, multi-functionality will be increasingly common in forest product systems. This development is spurred by the increased use of principles of industrial ecology and eco-efficiency when designing product systems (Heijungs and Guinée 2007).

In LCAs, multi-functionality becomes a problem when it is not feasible to split a multi-functional process into sub-processes connected to specific functions. Then, the LCA practitioner needs to find a rationale for allocating the environmental load of the multi-functional process between its functions, for example emission X between products A–C in Fig. 4.1. This can be handled by two principally different categories of methods, as acknowledged by ISO 14044 (ISO 2006b). The first category is *system expansion*, where the functional unit is redefined to account for the functions of *all* co-products (e.g. in the example of Fig. 4.1, the provision of products A, B and C), or a main product is selected among the co-products and given credit for the avoided environmental burdens from products assumed to be substituted by the other co-products (so-called substitution method). The second version of system expansion is commonly used although not acknowledged by ISO 14044, and has been criticised for introducing too much speculation in the system modelling (Heijungs and Guinée 2007). The second category of methods for handling multi-functionality is *partitioning*, whereby the environmental burden of the multi-functional process is divided between the co-products based on an attribute of theirs, for example a physical attribute such as mass, volume or energy content, or a monetary attribute such as production cost, market value or profitability.

The choice of method for handling multi-functionality can be based on recommendations in standards, guidelines or directives, such as ISO 14044 (ISO 2006b), the ILCD handbook (EC 2010) or the Fuel Quality Directive (EC 2009a). The choice of method can also depend on whether the study is attributional or consequential—partitioning methods are more common in attributional studies and system expansion is more common in consequential studies. Moreover, the choice of method can depend on preferences of the LCA practitioner or others influencing methodological choices (e.g. the commissioner of the LCA study), which in turn may (unfortunately) be influenced by some stakeholder's interest in "proving" the environmental benefits of a certain product or by strong beliefs regarding how multi-functionality ought to be handled (e.g. that a certain co-product should always be considered free of environmental burden). Preferences can also be expressed in terms of natural science approaches (allocation based on some physical attribute) or social/economic science approaches (allocation based on some economic attribute) (Pelletier et al. 2015). Ultimately, the choice of method can considerably influence the results of LCAs of forest products (Karlsson et al. 2014; Cherubini et al. 2011; Luo et al. 2009) and, as a consequence, decision-making based on those results. There has been critique that similar multi-functionality problems are often handled in different ways and that the choice of method is often poorly justified in LCA reports (Pelletier et al. 2015). Inconsistencies of practices may arise because of, for example, unclear or ambiguous recommendations in ISO 14044, or differences between recommendations in ISO 14044 and other standards (Weidema 2014).

4.2.1   Examples from the Literature

Cherubini et al. (2011) and Sandin et al. (2015c) provide clear examples of the influence of different methods for handling multi-functionality in LCAs of forest products. Cherubini et al. (2011) tested six methods: a substitution method as well as five partitioning methods (based on mass, energy, exergy, economic value as well as a hybrid partitioning method based on the environmental impact of a reference system with identical functions). The methods were applied in an LCA of a biorefinery system producing bioethanol, heat, electricity and phenols. Sandin et al. (2015c) also tested six methods: (i) allocation of all environmental burdens to the main product, (ii) substitution, (iii) exergy-based allocation, (iv) economic allocation, (v) the hybrid method proposed by Cherubini and colleagues, and (vi) allocation based on the co-products' potential for reducing environmental burdens if a reference system with identical functions is substituted (which was developed as an alternative version of the hybrid method). The methods were applied in an LCA of a biorefinery system producing paper pulp, a precursor to a vehicle fuel, methanol and heat; the methods were applied also on the forestry process, with the outputs pulpwood (the input to the biorefinery), timber, tops and branches, and fuelwood. Sandin and colleagues also evaluated the influence of the choice of method in four distinctly different decision contexts relevant for studies of biorefinery products. In both studies, the results varied considerably between the tested methods, which implies that the choice of method for handling multi-functionality is a critical methodological choice in LCAs of biorefinery products. The influence of the choice of method was particularly strong in consequential studies and in studies of physically non-dominant co-products. Results also showed that differences in the physical scale of co-product flows can determine the feasibility of applying certain methods. That the relative physical scale of co-product flows should influence the choice of method for handling multi-functionality and the need for comparing different methods is not recognised in LCA standards and guidelines such as ISO 14044 (ISO 2006b), the ILCD handbook (EC 2010) and the PEF guide (EC 2013). Recognising the role of the relative physical scale of co-product flows would improve standards' and other guiding documents' guidance on handling multi-functionality. This recommendation adds to a growing body of recommendations for improving existing standards' guidance on handling multi-functionality (see, e.g. Pelletier et al. 2015; Weidema 2014).

Additional findings of Cherubini et al. (2011) and Sandin et al. (2015c) relate to the investigated hybrid methods, which are potentially useful in consequential policy-related contexts which concern the mitigation of a certain environmental impact, where there are both material and immaterial co-products, where it is not feasible or desirable to choose a main product, and where it is desirable for results to be independent of fluctuating economic values. There are, however, some concerns and limitations of the hybrid methods. The method proposed by Cherubini et al. (2011) is potentially problematic because it favours products with a low potential to mitigate environmental burdens, and the results of the alternative hybrid method proposed by Sandin et al. (2015c)—which instead favours products with a high mitigation potential—were shown to be strongly influenced by the relative physical scale of the co-product flows. Thus, we recommend that LCA practitioners primarily use more traditional methods for handling multi-functionality, as recommended by the ISO 14044 standard.

4.3 Inventorying and Assessing Environmental Impacts

Established LCI and LCIA methods in LCA may fail to sufficiently address some environmental impacts of particular relevance for forestry and forest products. These impacts include but are not limited to climate change, biodiversity loss, disturbances to the water cycle and energy use. To improve the decision support provided by LCAs of forest products, there is a need to develop LCI and LCIA methods that can capture aspects of the environmental impact of forest products in a more relevant way than is done by established methods, and, until better methods are available, find ways of handling the shortcomings of established methods. Below sections describe some of the main challenges related to assessing climate, water cycle and biodiversity impacts as well as the energy use of forest products, with some examples of means for tackling the challenges. Before entering into the details of LCIA methods, a section outlines the challenges of modelling the carbon flows of the forest—which belongs to the goal and scope definition and LCI phases of an LCA rather than the LCIA phase. The modelling of carbon flows has direct influence on the assessment of the climate impact of forest products. Aspects of the modelling of carbon flows—temporal and spatial system boundaries and choice of baseline—are fundamental also in the modelling of other environmental impacts.

The challenges of assessing environmental impacts of forest products can be divided into the four categories listed below (adopted from Sandin et al. 2015a). Some of these challenges (i–iii) can be overcome with improved modelling, whereas others (iv) are rather about increasing the awareness among LCA practitioners in order to spur more context-aligned and transparent methodological choices.

(i)   Limitations in the understanding of how forests interact with other parts or aspects of natural systems (e.g. the climate).
(ii) Limitations in the understanding of how this interaction is influenced by the forest product system (particularly the extraction of forest biomass).
(iii) Limitations in the ability to model the aforementioned interactions and influences.
(iv) Value-based modelling choices, e.g. in terms of the setting of spatial and temporal system boundaries and the choice of baseline.

4.3.1 Modelling the Carbon Flows of Forest Product Systems

The description of challenges involved in carbon flow modelling is divided into three subsections, focussing on spatial modelling, temporal modelling, and choice of baseline. Most of these challenges relate to challenge (iv) above: value-based modelling choices. It should be noted that although a choice in LCA is value-based, it does not mean it is arbitrary—as always in LCA, methodological choices should be based on the goal and scope of the study and be clearly stated in the LCA report.

4.3.1.1   Spatial Modelling of Carbon Flows

In the world as a whole, biomass stocks in boreal and temperate forests are growing (Liski et al. 2003),[1] whereas stocks in tropical rain forests are decreasing (IPCC 2013). The biomass stocks may however be decreasing in certain temperate and boreal regions, and there may be constant or growing stocks of biomass in certain tropical regions. How to model forest biomass from certain boreal, temperate or tropical regions thus depends on the geographical location of the forest and the geographical resolution of the study. In other words, the modelling should depend on the spatial system boundaries. The choice of spatial system boundaries can be based on a global perspective, regional (e.g. the EU) or national borders, the forest ownership of a certain company, some other level at which forests are managed (e.g. a landscape perspective), or the specific stand that the harvested biomass is derived from (a single-stand perspective). In LCA, the choice of spatial system boundaries is often described as the choice between the latter two: the landscape

---

[1]Swedish forest biomass stocks doubled in 1926–2008 (Swedish University of Agricultural Sciences 2011).

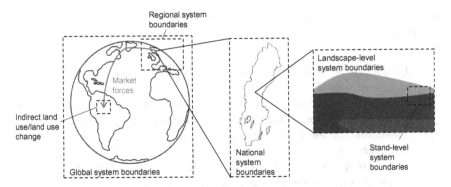

Fig. 4.2 Visualisation of spatial aspects influencing carbon footprints of forest products, with permission reproduced from Sandin et al. (2015a)

and the single-stand perspectives. A single-stand perspective means accounting for the re-growth of biomass on the same stand as the harvested biomass, while a landscape perspective means considering the forest in a larger area—including different age classes—as a unit and accounting for net biomass increase or decrease of this unit (Cherubini et al. 2013). The choice of perspective should depend on the context of the study—e.g. on the responsibilities of the commissioner of the study and the specific decisions the study is intended to support—and can considerably influence how one views the renewability and carbon neutrality of the harvested forest biomass (Sandin et al. 2015a).

The inclusion or exclusion of indirect land use and land use change, as described in Sect. 2.5, is also a matter of spatial system boundaries.

Figure 4.2 illustrates the important aspects of the spatial modelling of forest carbon flows.

4.3.1.2   Temporal Modelling of Carbon Flows

An important aspect of the temporal modelling of carbon flows is the question of *when* the biomass growth is assumed to occur. With a landscape (or higher) perspective, one can assume that the biomass growth occurs the same year as the harvest—i.e. at the national, regional or landscape level, the annual biomass growth is allocated to the annual biomass harvest. However, with a single-stand perspective, one can either assume that the biomass growth occurs *before* the harvest (i.e. that one considers the growth of the biomass that will actually be in the studied product), or *after* the harvest (i.e. on consider the growth that will occur as a consequence of the harvest operation). The choice between these two perspectives can, for example, depend on whether an attributional or consequential perspective is applied. An argument for the first assumption can be that the trees were planted with the purpose of harvesting, and an argument for the second assumption can be that replantation is a consequence of harvesting, as this is often required by legislation or forestry certification schemes.

Another aspect of the temporal modelling of carbon flows is the timing of flows: CO_2 capture in the forest as well as fossil and non-fossil GHG emissions. Typically, timing is disregarded, so the climate impact of an emission pulse is assumed to be the same regardless of when it occurs. This is potentially problematic for two reasons. First, if using GWP_{100} (further described in Sect. 4.3.2), the 100-year time period is not applied consistently: for emissions occurring today, the metric includes impact within 100 years from today; but for future emissions, it includes impact within 100 years from the moment these emissions occur. Secondly, the risk of passing critical tipping points in the climate system implies that urgent impact mitigation is necessary and that it matters greatly whether emissions occur now or in, for example, 50 years (Jørgensen et al. 2014; Helin et al. 2013; Levasseur et al. 2010). Ignoring the timing is particularly problematic for forest products because GHG emissions (if forest biomass is used for long-lived products such as constructions) and CO_2 capture (as forests are relatively slow-growing) often occur over a long time period. This was emphasised in a recent study on how accounting for the timing of emissions influences the GWP scores of 4034 LCI datasets in the ecoinvent database, which found that GWP scores of datasets associated with the wood sector are particularly sensitive to assumptions related to the timing of emissions (Pinsonnault et al. 2014).

4.3.1.3 Choice of Baseline

Another important aspect in the modelling of carbon flows of forest product systems is what one assumes happens in the forest in the absence of the harvest, i.e. the so-called baseline or reference/counter-factual situation. Defining a baseline is needed to separate the technosphere from the natural system (Soimakallio et al. 2015) and to quantify the LCI data belonging to the product system. A baseline is needed both for carbon flow modelling (as a basis for climate impact assessment) and for modelling other environmental interventions, such as the biodiversity impact of land use (Lindqvist et al. 2015) and water cycle disturbances; in this subsection, the focus is on carbon flow modelling. Figure 4.3 provides three examples of possible choices of baselines in carbon flow modelling and illustrates that the choice of baseline can significantly influence the amount of CO_2 allocated to the harvested biomass. The upper part of the figure show the consequences of the choice of baseline for models based on a single-stand perspective, whereas the lower part of the figure shows possible outcomes for models based on spatial system boundaries derived from a landscape (or higher) perspective (assuming annual biomass growth within the system boundaries).

Reference situation 1 in Fig. 4.3 allocates all CO_2 captured in the forest to the harvested biomass, for example based on the assumption that there are no net carbon flows in the absence of harvest because the forest has reached a steady-state. Reference situation 2 allocates the difference between the CO_2 captured in the

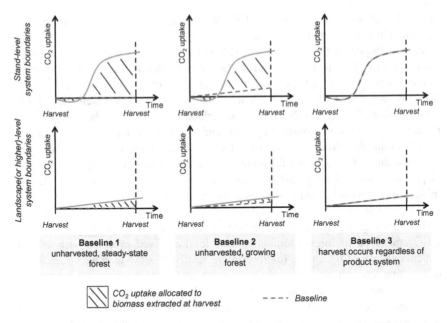

Fig. 4.3 Possible baselines in determining which part of the carbon flows that should be allocated to the studied product system, with permission reproduced from Sandin et al. (2015a). The curves are produced solely for illustrating potential consequences of the choice of baseline, and are not based on real data

harvested forest, and the CO_2 captured in an unharvested but growing forest, to the harvested biomass; Koponen and Soimakallio (2015) describe this as "using natural regeneration as the land use baseline". Reference situation 3 implies that harvest is assumed to occur regardless of the product system.

It has been argued that attributional studies should not account for baselines, as these are hypothetical, thus belonging to a consequential tradition. However, the quantification of LCI data always requires a baseline. When modelling carbon flows in the forest, ignoring baseline most likely means that one implicitly assumes reference situation 1 in Fig. 4.3. In contrast, Soimakallio et al. (2015) argue that the most coherent baseline in attributional studies would be reference situation 2.

We recommend that reference situation 3 is not applied in LCA, because assuming that the harvest occurs regardless of the product system is equivalent to assuming that also other natural resources, such as fossil resources, would be used regardless of the product system. Such a perspective entails that no environmental stressors should be allocated to the studied product system—not a very constructive perspective in studies supposedly used to support decisions aiming for reduced environmental impacts (which ought to be the ultimate aim of any LCA-based decision making).

Until recently, the choice of baseline has not received proper attention in the scientific literature on the modelling of carbon flows in LCA. Hopefully, the contributions from Soimakallio et al. (2015), Koponen and Soimakallio (2015) and Brander (2015), will initiate a discussion on the topic that can lead to more conscious handling of the baseline in LCAs of forest products.

4.3.2 Climate Change

Difficulties in assessing the climate impact of forest products mainly relate to temporal dynamics of carbon flows in the forest between soil, vegetation and air, from sowing to harvest and regrowth (as touched upon in the previous subsection), temporary storage of carbon in the product life cycle, and non-carbon climate effects such as changes in the albedo. To understand these difficulties, there is a need to introduce how the climate impact of forest products is usually assessed in LCAs.

The currently most common metric to assess the climate impact of products is the GWP, which is an indicator of how much a GHG emitted to the atmosphere influences the radiative forcing under a set time period. Radiative forcing is a measure of the balance between the incoming solar radiation and the energy radiated back to space (IPCC 2013). Since different GHGs have different atmospheric residence times, the chosen time period influences the relative impact of different types of GHGs. In LCA, it is usual to use a time period of 100 years (indicated by a subscript: GWP_{100}). This means that in comparing the contribution of different types of GHGs, one considers climate impact occurring within 100 years—an arbitrary choice (Reap et al. 2008) often made without a clear motivation (Røyne et al. 2016). Furthermore, biogenic CO_2 emissions are most often considered to have a GWP of zero, i.e. they are assumed to be climate neutral (Røyne et al. 2016). The climate neutrality of biogenic CO_2 is based on the assumption that forest products (and likewise other bio-based products) are carbon neutral, i.e. there is a balance between carbon sequestered in the forest during growth and carbon emitted to the atmosphere once the forest biomass is incinerated (which may be in the product's end-of-life phase, or in case of material recovery: in the end-of-life phase of some subsequent product). As concluded in Sect. 4.3.1, whether it is valid or not to assume carbon neutrality in a specific study depends on the spatial and temporal scope of the study and on other methodological choices (e.g. whether or not the study accounts for indirect land use and land use change). Even in cases where it is valid to assume *carbon* neutrality, assuming *climate* neutrality may not be valid, as a temporal shift between emitted and sequestered carbon may temporarily contribute to a changed radiative forcing (Helin et al. 2013). How to consider this temporal shift depends, among others, on whether one attributes the growth before or after the harvest to the product system under study (as discussed in Sect. 4.3.1). It has been shown that results of LCAs of forest products can depend strongly on whether or not biogenic CO_2 emissions are considered climate neutral (Røyne et al. 2016; Sandin et al. 2015a; Garcia and Freire 2014; Guest and Strømman 2014; Zanchi et al. 2012; Sjølie and Solberg 2011), and assuming climate neutrality by default has therefore been questioned (Klein et al. 2015; Ter-Mikaelian et al. 2015; Agostini et al. 2013; Gunn et al. 2012; Schulze et al. 2012; Johnson 2009; Searchinger et al. 2008). It should be noted that abandoning the climate neutrality assumption does not necessarily imply an increase in the calculated climate impact of forest products, as a well-managed forest can function as a net carbon sink and thus result in a negative (i.e. beneficial) climate impact (Guest and Strømman 2014; Perez-Garcia et al. 2005; Liski et al. 2003).

Another often ignored aspect of the climate impact of forest products, which relates to the timing of CO_2 capture and GHG emissions, is the temporary storage of carbon in forest products. Assuming that the harvesting of this carbon (in the form of forest biomass) allows more CO_2 to be captured in the forest than would otherwise have been the case, and assuming that the storage of this carbon in products prevents the carbon from being emitted to the atmosphere for some time, it has been argued that carbon storage in products causes a temporary reduction in radiative forcing which should be accounted for in the climate impact assessment of forest products (Vogtländer et al. 2014; Brandão et al. 2013; Levasseur et al. 2010; Costa and Wilson 2000). On the other hand, research has also concluded that temporary carbon storage may not reduce climate impact because it lowers the CO_2 gradient between the atmosphere and potential carbon reservoirs, such as the oceans, and thus reduces CO_2 removal from the atmosphere. Once the temporarily stored carbon is released again, the atmospheric CO_2 concentration is, the argument goes, higher than would be the case without temporary storage (Kirschbaum 2006). This way of considering temporary carbon storage has, however, been criticised because it neglects the cumulative climate impact (Dornburg and Marland 2008).

Other potentially important aspects of climate impact, which are seldom captured in LCAs of forest products, are the climate impact due to disturbances of soil organic carbon (Brandão et al. 2011; Repo et al. 2011; Stephenson et al. 2010) and other disturbances of soil biogeochemistry causing emissions of nitrous dioxide (N_2O) and methane (CH_4) (Repo et al. 2011; Wrage et al. 2005; Cai et al. 2001). There are some further seldom-captured aspects of the climate impact of land use and land use change unrelated to GHGs emissions that are potentially important in LCAs of forest products. For instance, the albedo of the Earth's surface—i.e. its capacity to reflect sunlight back into space—increases when the surface is bright (e.g. snow-covered) and smooth (e.g. a clear-cut forest), factors which may be influenced by land use and land use change (Cherubini et al. 2012; Schwaiger and Bird 2010). Changes in the reflection of solar radiation may also be induced by land use and land use changes that impact the forest's ability to form organic vapours. These vapours can create aerosols that reflect sunlight or form particles which catalyse the formation of clouds, which in turn reflect sunlight (Spracklen et al. 2008).

As indicated above, there are many potentially important aspects of the climate impact of forest products that are seldom accounted for by established impact assessment methods and practices. There are, however, new methods available that are capable of capturing some of these aspects. So far, the focus of the development of new methods has been on capturing the temporal dynamics of the forest product systems: timing of CO_2 uptake and GHG emissions, carbon storage in the product, etc. Examples of such methods are GWP_{bio} (Guest et al. 2013) and Dynamic LCA (Levasseur et al. 2010). The latter method has recently been tested in combination with various spatial system boundaries (based on national, landscape and single-stand perspective) for five different forest product systems: two vehicle fuels, a building component, textile fibres, and an industrial chemical (Sandin et al. 2015a). In the study it was concluded, among others, that a temporally more advanced climate impact assessment method (compared to the traditional GWP_{100} metric and how it is usually applied in LCA) is necessary to capture the nuances in the climate impact of different uses of forest biomass.

The differences between LCIA results derived from using currently common versus alternative impact assessment practices were also recently mapped by Røyne et al. (2016). Results showed that the current methods and practices exclude most dynamic features of carbon uptake and storage as well as climate impact of indirect land use change, aerosols and changed albedo. Results also showed that including these aspects can strongly influence LCIA results, both positively and negatively, with potentially important implications for decision-making (four specific decision-making contexts were studied). For example, the elements of the product life cycle that have the greatest climate-impact reduction potential might not be identified, product comparisons might favour the product with highest climate impact, and policy instruments might support the development and use of inefficient climate-mitigation strategies. Also, the study identified a limitation with several of the non-established methods: the difficulty of finding the location-dependent data required by the methods (e.g. information on forestry practices and soil conditions, or information on regional weather patterns influencing the albedo effect). Furthermore, the study concluded that in contexts dealing with short-term climate impact mitigation, it is preferable to use climate impact metrics based on a shorter time perspective (e.g. GWP_{20}), or to use dynamic metrics or to discount future emissions—options which are not among the established methods and practices. In such contexts, if the harvested forest is slow-growing, it could also be suitable to account for temporary carbon storage (e.g. in long-lived products) and the climate impact of the temporary contribution of biogenic CO_2 emissions to radiative forcing. Moreover, short-term mitigation implies that albedo effects and soil disturbances could be relevant to account for as long as these are short-term effects. End-of-life substitution credits are in such contexts probably less important to consider, particularly in studies of long-lived products. Also in contexts dealing with long-term climate impact mitigation, temporally more advanced methods could be suitable (e.g. such as GWP_{bio} or dynamic LCA). However, in contrast to contexts dealing with short-term mitigation, metrics based on a longer time perspective (e.g. GWP_{500}) are preferable. Also, in such contexts, it could be justifiable to disregard biogenic CO_2 emissions (if it can be reasonably assumed that the harvested biomass is carbon-neutral in the long term), soil disturbances (if these can be assumed to be reversed over time) and the albedo effect (unless harvesting results in permanent deforestation). It thus appears that the current impact assessment methods and practices, which do not consider most of the aforementioned climate impact aspects, are more suitable for contexts dealing with long-term impact mitigation, than for contexts dealing with short-term impact mitigation.

The aforementioned potential shortcomings of established climate impact assessment methods and practices warrant the development of more sophisticated methods and practices that consider more of the climate impact aspects of forest product life cycles, and the inclusion and consideration of such methods and practices in important standards, guidelines and directives, such as ISO 14040/14044 (ISO 2006a, b), the ILCD handbook (EC 2010), the Renewable Energy Directive (EC 2009b), and the PEF recommendations (EC 2013). Omitting climate impact aspects known to be important from impact assessment modelling is simply not acceptable—or, in the words by Sterman (1991, p. 12) on system modelling in general: "ignoring a relationship implies that it has a value of zero—probably the only value known to be wrong".

Røyne et al. (2016) also showed that methods and practices are seldom clearly stated and/or motivated in published studies, and often do not seem to have been chosen based on the specific context of the study. This calls for improved practices of LCA practitioners both in terms of more context-adapted methodological choices and in terms of being transparent about the choices made.

Improved methods and practices can ensure more accurate assessments of the climate impact of forest products and ensure the avoidance of over-simplified and untrue statements regarding the benefits of forest products vis-à-vis non-forest products. This will be essential if we are to use forest and non-forest resources efficiently for mitigating climate change.

4.3.3   Biodiversity Loss

The development of improved methods for assessing biodiversity loss is most often part of a broader development of impact assessment methods for land use impacts, which is therefore the starting-point for the discussion in this section. Traditionally in LCA, simple proxy indicators have been used to account for the environmental impact of land activities—for land use: the land area used and the duration of the use (e.g. in m^2 * years), and for land use change: the land area transformed (e.g. in m^2) from one state (e.g. forest) to another (e.g. agriculture). However, research has concluded that these are meaningless indicators, as land use and land use change can cause both positive and negative impacts, depending on the specific location of the land activities (Michelsen et al. 2012). As a consequence, there have been many proposals for new impact assessment methods, with indicators reflecting a more sophisticated characterisation of the environmental impact of land use and land use change. The development of new methods is facilitated by an increasing understanding of the ecosystem services provided by functioning land ecosystems and the increasingly harmonised classification of such services (Haines-Young and Potschin 2013; Fisher et al. 2009; MA 2005). The development is also facilitated by the increased availability of databases on land use and land cover, such as the

GlobCover database on global land cover (European Space Agency 2011), the Terrestrial Ecoregions of the World (TEOW) classification (Olson et al. 2001) and the Corine database on European land cover (European Environment Agency 1995). Moreover, key for this development is the UNEP/SETAC life cycle initiative (Koellner and Geyer 2013), which leads a consensus process for agreeing on principles for inventorying and characterising environmental impacts (primarily biodiversity impact) of land use and land use change anywhere on Earth (Teixeira et al. 2015; Koellner et al. 2013a, b). This initiative constitutes an important step in harmonising the plentiful recent advancements (reviewed below) of land use impact assessment methodology.

Proposed impact assessment methods often focus on biodiversity impact, as reviewed by Curran et al. (2011) and De Souza et al. (2015). This makes sense since biodiversity loss is a major reason for weakened ecosystem services (Gamfeldt et al. 2013; Hooper et al. 2012; Thompson 2011; MA 2005; Chapin et al. 2000). Among biodiversity indicators, species richness of a certain group of species has been a commonly used indicator. Primarily, the focus has been on the species richness of vascular plants (Schmidt 2008; Goedkoop and Spriensma 2000; Koellner 2000; Lindeijer 2000), but there have been proposals to consider other species, alone or in combination. Proposed indicators include the richness, abundance and evenness of vertebrate species (Geyer et al. 2010), the combined species richness of vascular plants, molluscs, mosses and threatened species (Koellner and Schulz 2008) and the combined species richness of vascular plants, birds, mammals and butterflies (Mattsson et al. 2000). de Baan et al. (2012) proposed a method using the relative species richness as an indicator, which allows the use of data on any group or groups of species for which species richness data exist.

Other methods for assessing impacts of land use and land use change measure other facets of biodiversity or other ecosystem attributes (which may indirectly influence biodiversity), sometimes in combination with species richness indicators. Examples include the measurement of functional diversity (species richness data combined with functional traits[2] data; De Souza et al. 2013), global potential for endemic species extinction (de Baan et al. 2013), net primary productivity (Weidema and Lindeijer 2001; Lindeijer 2000), biotic production potential (Brandão and Milà i Canals 2013), naturalness as defined by 11 qualitatively described land classes (Brentrup et al. 2002), soil quality (Saad et al. 2011; Milà i Canals et al. 2007; Mattsson et al. 2000), soil erosion (Núñez et al. 2013), number of red-listed species in combination with several biotope-specific key features (Kylärkorpi et al. 2005), and conditions for maintained biodiversity (based on the amount of decaying wood, the area set aside and the introduction of alien species; Michelsen 2008). Many of these methods utilise several indicators. For example, the biotic production potential method by Brandão and Milà i Canals (2013) is based on several indicators of soil quality. Moreover, several of the mentioned

---

[2]The functional traits of a species are "morpho-physiophenological traits which impact fitness indirectly via their effects on growth, reproduction and survival" (Violle et al. 2007, p. 1).

publications advocate the use of other complementary methods for achieving a holistic view of the impact on ecosystem quality. There have also been attempts to combine indicators of different ecosystem attributes into a single index. Such attempts include a procedure for combining 18 indicators to measure the exergy of ecosystems [i.e. the energy available for work in an ecosystem, which (it is argued) reflects the ecosystem quality; Muys and Quijano 2002], and a methodological framework for combining several biodiversity-related parameters at the level of regional ecosystems into a regional biodiversity potential metric (Lindner et al. 2014).

Sandin et al. (2013) tested the method proposed by Schmidt (2008) in an LCA case study that compared regenerated cellulosic textile fibres from forest biomass with cotton fibres. The method uses the species richness of vascular plants as a proxy for biodiversity and distinguishes between impact of land use (termed occupational impact) and impact of land use change (termed transformational impact). The method enables the calculation of characterisation factors that depend on the geographical location of the land, accounting for its altitude and latitude and the intensity of land use in the surrounding region—factors that influence the vulnerability of ecosystems to interventions such as land use and land use change. The possibility to calculate location-dependent characterisation factors is important for distinguishing the influence of the location of operations. Most other proposed methods for assessing the biodiversity impact of land use and land use change were developed for the assessment of land activities in specific regions, and, due to a lack of data, they are typically not applicable in assessments of globally distributed supply chains—which is increasingly common in forest product life cycles.

Sandin et al. (2013) found that (in the specific case study) transformation of land from a high- to a low-biodiversity state contributed much more to the biodiversity impact than occupational land use. Moreover, the location of land use was shown to be of low importance, as geographical differences influencing the time from planting to harvest, the annual yield per land area, the renaturalisation time and the ecosystem vulnerability appeared to roughly offset each other (this may not be the case in other case studies). Thus, the method by Schmidt (2008) enabled Sandin and colleagues to identify important product system parameters beyond what would have been possible by using less sophisticated methods (such as the simple indicator on land use described in Sect. 4.3: the area of land used and the duration of that use). Although more sophisticated methods are available (of which the method by Schmidt is just one example), there is still a lack of land use impact assessment methods that can assess differences between closely related land uses (such as different practices in forestry management). Mainly this is because of limitations in data availability. Therefore, methods presently available in the scientific literature are particularly useful for assessments supporting strategic macro-scale decision-making (e.g. whether to transform natural land or not, or which regions to source forest biomass from), but less useful for supporting micro-scale decision-making (e.g. what specific forest to source biomass from, or what land management practices to use). This drawback can, however, be overcome with more refined data, such as species richness data for more specific land management practices, which could facilitate comparisons between uncertified land and land managed according to certain certification principles, such as the Forest Stewardship Council (FSC 2016) or the Programme for the Endorsement of Forest Certification (PEFC 2016).

Sandin et al. (2013) also identified another critical methodological aspect of land use impact assessment: the allocation of transformational impacts between the first harvest after transformation and subsequent harvests (for more about allocation problems, see Sect. 4.2). This proved to be very important in comparing regenerated cellulose fibres made from biomass from forests with a rotation time of 62.5 years and cotton fibres from cotton plantations with a rotation time of 0.5 years. This allocation problem can be expected to be a recurring dilemma of importance for many comparisons of products derived from crops with different rotation periods, for example in comparisons of forest products and other bio-based products, or in comparisons of forest biomass with different rotation periods. Indeed, the problem was recognised by Koellner et al. (2013b), where it was recommended that, as a base case in the absence of a scientifically robust alternative, the transformational impact should be allocated over 20 years regardless of crop. They also recommended testing the influence of other amortisation periods in a sensitivity analysis. Furthermore, Schmidt et al. (2015) recently proposed an approach for avoiding the allocation of transformational impacts between harvests; the proposed approach is, however, only applicable for the assessment of climate impact.

A further conclusion by Sandin et al. (2013) was that the assessment method for biodiversity loss could be improved and/or complemented by other indicators, on other groups of species or other facets of biodiversity and/or ecosystem quality. The need for multiple indicators for the assessment of land use activities has been emphasised elsewhere as well (Teixeira et al. 2015; Koellner et al. 2013b; Curran et al. 2011). As seen in the above review of methods, some of the proposed methods are based on multiple indicators. The method suggested by de Baan et al. (2012), using a relative species index, is perhaps the most promising development in this direction. Their method permits using several groups of species, where the choice of species groups can be based on the data available for a certain region. However, it does not yet support transformational impact assessment, which, as shown by Sandin et al. (2013), needs to be included in a robust method for assessing the environmental impacts of land use and land use change.

Finally, it should be said that the general principles for LCI and LCIA methods for environmental impacts of land use and land use change proposed by the UNEP-SETAC life cycle initiative (Koellner et al. 2013a, b), could prove to be an important step towards finding a consensus in the LCA community. The proposed principles are largely consistent with the method by Schmidt (2008): they emphasise the need to distinguish between transformational and occupational impacts, recommend the use of reference states (or "baselines", see Sect. 4.3.1) to identify changes in ecosystem quality, acknowledge loss of species diversity as an important link in the cause-effect chain between land activities and the areas of protection, recommend the use of absolute metrics for biodiversity loss (in contrast to relative metrics, such as the percentage of species lost within a given area), emphasise the need for geographical differentiation of characterisation factors, and acknowledge the importance of regeneration time and its geographical dependence.

4.3.4 Water Cycle Disturbances

The traditional inventory-based proxy for assessing water cycle disturbances (or water use impact, as it is traditionally referred to; the terminology is further discussed in Sect. 4.3.6)—i.e. the volume of freshwater used—has been criticised for not correlating with actual impacts further down the cause-effect chain (Ridoutt et al. 2012; Ridoutt 2011). This has led to intense development of more elaborate methods, as reviewed by Berger and Finkbeiner (2010), Kounina et al. (2013) and Boulay et al. (2015a). These reviews were done within the water use in LCA (WULCA) working group in the UNEP-SETAC life cycle initiative, which has initiated a process for finding a consensus on the assessment of water cycle disturbances, similarly to the initiative for finding a consensus on land use impact assessment, as mentioned above (WULCA 2014). In parallel, there is another consensus process that has recently resulted in an international standard for water footprinting, ISO 14046 (ISO 2014). In the development of more elaborate methods for assessing water cycle disturbances, two main difficulties arise: what volume of water to consider in the LCI (i.e. quantifying alterations to the water cycle) and how to interpret this volume in terms of environmental impact in the LCIA.

In LCIs of bio-based products, apart from including process water, it has been common to include engineered water supplied to the crop, for example by irrigation systems, and to disregard naturally supplied water, for example from precipitation (Peters et al. 2010). More elaborate LCI approaches have been developed, which consider the water use of the metabolism of the crop by attributing evapotranspirational losses to the studied product—approaches which also account for the use of naturally supplied water, as such water use may influence the water cycle in terms of water availability downstream and thus have environmental consequences (see, e.g. Hoekstra and Chapagain 2007). Attributing evapotranspirational losses to forestry and forest products has, however, been questioned as such losses occur also in unmanaged forests (Launiainen et al. 2014). Moreover, LCI methods for water use generally disregard catchment-scale effects of land use and land use change, for example effects on water runoff due to factors such as the interception of rainfall by vegetation, forestry road construction or changes in soil structure (Bruijnzeel 2004; Swank et al. 2001). These factors may be irrelevant when land use is considered a static system dominated by monocultures, but they certainly are relevant for more complex land use systems, such as forestry, and in cases of land use change. Therefore, Sandin et al. (2013) suggested a consequential LCI approach that attempts to capture such factors and avoid relying on evapotranspirational losses as the basis for quantifying water cycle alterations. The approach accounts for the change in water runoff that occurs as a result of forestry during harvesting and the subsequent regrowth of trees. This captures not only the water demand by the harvested trees, but also the total influence on downstream water availability by forestry operations. Changes in runoff have previously been suggested as an important impact pathway for water cycle disturbances (Milà i Canals et al. 2009; Heuvelmans et al. 2005) and have been included in LCIs of water use for static agricultural systems (Peters et al. 2010). However, Sandin et al. (2013) provides the first example of including changes in runoff in LCIs for systems with forestry or land use change [recently, however, changes in runoff have been accounted for in an LCA of a forestry system without land use change (Quinteiro et al. 2015)]. Sandin et al. (2013) compared the consequential LCI approach with a more traditional attributional LCI approach, based on the evapotranspirational losses of the harvested trees. Figure 4.4 illustrates the difference between the attributional and consequential approaches. The development of this consequential LCI approach is one example of the many parallel ongoing developments regarding how to improve the inventorying of water use in LCAs of forest products.

There are many suggestions for how to characterise the environmental impact of the water volume quantified in the LCI, i.e. the impact caused by the water cycle alteration. Bösch et al. (2007) proposed an exergy indicator for resource consumption, including the use of water. Frischknecht et al. (2009b) proposed a method for characterising water use within the ecological scarcity LCIA framework, in which the water volume withdrawn from a region (or consumed) is

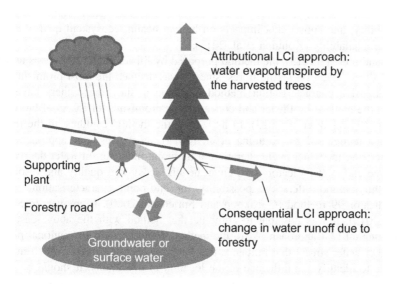

Fig. 4.4 Schematic view of water flows in the forestry system. With an attributional LCI approach, the forestry's alteration of the water cycle is estimated by the evapotranspirational losses of the trees during their growth. With the consequential LCI approach, the alteration of the water cycle instead refers to the change in runoff that occurs as a result of forestry during harvesting and the subsequent regrowth of trees, which captures the influence of factors such as the construction of forestry roads and the planting of supporting vegetation

characterised based on the ratio between the total water volume withdrawn from the region and the critical water use volume for the region (set at 20 % of the renewable water available in the region). Milà i Canals et al. (2009) discussed a number of impact pathways of how freshwater use and land use change may lead to freshwater stress and subsequent impacts on human health and ecosystem quality, and how the use of fossil and aquifer groundwater may reduce freshwater availability for future generations. In an assessment of "freshwater deprivation for human uses", Bayart et al. (2009) distinguished between the quality (low or high) and the type of water (surface water or groundwater) entering and exiting the studied system. Motoshita et al. (2008, 2009) proposed methods for assessing undernourishment-related human health damages due to agricultural water scarcity, and of human health impacts due to infectious diseases originating from domestic water use. Van Zelm et al. (2009) proposed a method for assessing the ecosystem quality impact of groundwater extraction, specific for Dutch conditions. The WFN proposed a method for aggregating different types of environmental impacts related to water, for example impacts of water use and impacts of water pollution (Hoekstra et al. 2011). As a proxy for the pollution impacts, the WFN method uses the water volume necessary to dilute emissions to freshwater to such an extent that the water quality adheres to water quality standards. Other such single-score approaches have also been suggested, drawing on the latest developments in the LCIA modelling of water-related environmental impacts (Bayart et al. 2014; Ridoutt and Pfister 2013). Most suggestions for characterisation methods in some way relate impacts of water use to water scarcity, water functionality, water ecological value or water renewability, and subsequent impacts on human health, ecosystem quality and/or resource availability (Kounina et al. 2013).

In Sandin et al. (2013), the method proposed by Pfister et al. (2009) was used in the LCA case study on textile fibres, since it was deemed the most promising and comprehensive characterisation method available at the time. Others have also deemed the method by Pfister and colleagues as promising for LCAs of bio-based products (Pawelzik et al. 2013). In its favour, the method captures all the impact pathways recognised by Kounina et al. (2013), as it uses four approaches for characterising the water cycle alteration: a midpoint indicator on water deprivation and three endpoint indicators on human health, ecosystem quality and resources. Also, the method offers the possibility for end-point characterisation by the eco-indicator 99 method (Goedkoop and Spriensma 2000). Moreover, in Sandin et al. (2013), it was possible to combine the method with the above-described consequential LCI approach. Also, the method could account for regional parameters (e.g. water stress) that influence the water cycle disturbances, which made it possible to identify the influence of the location of operations. It should be noted that the method offers the possibility to define characterisation factors at an even finer resolution than done in Sandin et al. (2013).

Sandin et al. (2013) showed that the location of operations is critical for the results, as water extracted from relatively water-stressed environments cause higher impacts. This is an obvious advancement compared to the most commonly used impact assessment methods in LCA, which simply give the volume of water used, sometimes distinguishing between surface water and groundwater, but disregarding whether this represents an actual alteration of the water cycle and not acknowledging the potential consequences further down the cause-effect chain.

Furthermore, some results in Sandin et al. (2013) differed considerably between the novel consequential LCI approach and a traditional attributional LCI approach. In particular, it was shown that the consequential approach offers the possibility to calculate catchment-scale effects of land use and land use change not possible with previously proposed LCI approaches, such as the possibility of recognising increased runoff as a potential benefit for certain types of land use—a development of the impact assessment which reflects realities in a meaningful way. Apart from potentially providing a more accurate quantification of water cycle alterations, the consequential LCI approach could offer further opportunities if combined with more elaborate LCIA methods. For example, a smaller time step in the estimation of runoff change could theoretically make it possible to account for the fact that increased water runoff may be a potential *disadvantage* in some regions, by expanding the LCIA modelling with an index on the sensitivity to flooding, erosion, soil salinity or similar. It has even been argued that the beneficial impacts of increased runoff due to land use change should never be accounted for unless damaging effects are also considered (Berger and Finkbeiner 2012). Apart from the aforementioned damaging effects due to increased runoff, Berger and Finkbeiner acknowledged other potentially damaging effects from land use change, such as decreased precipitation in other catchments. Including effects such as flooding would be a considerable development of established methods for assessing water cycle disturbances—which solely focus on water deficiency as a potential issue. The desirability of such advancement is the primary reason for why the impact category is, in the present book, referred to as "water cycle disturbances" rather than the more traditional term "water use impact" or some term referring exclusively to water deficiency (the terminology is further discussed in Sect. 4.3.6).

When applying the consequential LCI approach for quantifying water cycle, Sandin et al. (2013) made some rough estimates of runoff changes. These estimates are very uncertain; for example, estimates made for Swedish forestry resulted in an interval spanning from a negative to a positive change in runoff. Since the publication of Sandin et al. (2013), others have argued that Scandinavian forestry (which includes Swedish forestry) has only a marginal influence on runoff (Launiainen et al. 2014). Thus, to increase the usefulness of the consequential LCI approach, there is a need for more accurate and abundant data on runoff changes for different land use types and management regimes. To achieve this, there is a need to discuss the spatial and temporal representativeness of available data, and collect new data when necessary. Without high-quality data, there is a risk that increasing sophistication will add more noise than information (Hertwich and Hammitt 2001), and will fail to contribute to improved decision support.

While work by Sandin et al. (2013), Peters et al. (2010) and others have focussed predominantly on questions of life cycle inventory as a prerequisite for assessing water use impacts, the WULCA group has applied more effort to the area of LCIA indicator development. A series of international workshops has been run to try and generate consensus around the best way forward for these calculations (Boulay et al. 2015b). Three approaches were shortlisted: DTA, DTAx and AWaRe. DTA stands for "demand to availability", in other words: the water demanded by a system is divided by the available water. Of the three options, this was eliminated first on the basis that it had a low discriminatory power (watersheds tended to either have plenty of water or hardly any). To rectify this, WULCA offered a modification which saw the DTA scaled by the area of land associated with the available water ("DTAx"), which had the effect of more evenly spreading the stress indicator results for watersheds over two orders of magnitude. The current consensus is however that AWaRe (the inverse of availability minus demand, normalised to the weighted global average of this value) is superior for its discriminatory power and for being more physically meaningful than DTAx. Availability is adjusted for existing human and ecosystem demands and is calculated relative to the land area at the sub-catchment level (and subsequently aggregated as necessary). At the time of writing, the AWaRe method has not yet been evaluated in case studies.

Ten years ago, LCA analysts had no established proxies for water use impact that were better than simply reporting water use in litres. Because of the afore-mentioned method developments and international initiatives, we expect the level of interest in this field to continue to be high and are optimistic about the prospects for a global consensus.

4.3.5   Energy Use

Energy use is one of the oldest and most widely applied indicators in environmental assessments. It is sometimes used as a proxy for environmental impacts in general and sometimes as an indicator of energy resource use alongside other indicators (see Huijbregts et al. 2010 for a discussion on its suitability as a proxy indicator). The fact that this indicator is so common makes it important to discuss also in the context of forest products. Furthermore, as will be elaborated on below, as extracted forest biomass contains energy and is sometimes even exploited for this very reason, results from energy use assessments will be strongly affected by whether such energy is considered or not and how it is quantified.

Many different energy use indicators have been applied in literature but there is a large variation and also inconsistencies in how these are termed and how they are applied. Unfortunately, communication around these matters is also often insufficient. Together, poor communication of the scope of applied indicators and inconsistencies in terminology make it challenging to interpret results found in literature. Based on a review of energy assessments and LCAs of biofuels, Arvidsson and Svanström (2016) concluded that differences in how energy use indicators are applied mainly relate to (i) whether only fossil or also renewable energy is included, (ii) whether only flows that are intended for use as energy commodities or also flows that are intended for other purposes (such as process chemicals or materials eventually making up a physical product) are included, and

(iii) whether flows are accounted for at the level of primary energy or at the level of secondary energy. In two different studies, Arvidsson and collaborators showed that the application of five different common energy use indicators in case studies on the biofuel palm oil methyl ester (Arvidsson et al. 2012), and cellulose nanofibrils (Arvidsson et al. 2015a), respectively, yielded very different results and argued that the choice of energy use indicator is an important one that should be active and be based on the needs of the specific study, and this should be communicated clearly and transparently. To provide guidance, Arvidsson and Svanström (2016) developed a framework to be used in the design and communication of studies on energy use of different products. The framework focuses on the three different aspects that were found to be the main differences between energy use indicators, listed earlier in this paragraph (i–iii). They also categorized commonly used indicators according to the framework.

Cumulative energy demand (CED), for example, is a common energy use indicator, and it has been shown to be a good proxy indicator for other types of environmental impacts (Huijbregts et al. 2010). It is among the indicators that will yield high values in a comparison between energy use indicators as it includes both renewable energy and also extracted material that is intended for other purposes than energy purposes, and, it is reported on a primary energy level, i.e. including flows that are lost as losses in different transformation processes. Most other indicators would yield lower values as they would either disregard some flows or count only parts of the flows that become useful at some level. Only indicators that take into account even more of the extracted material would yield higher values. But, how much of the forest biomass that is moved away from the site at harvesting or even just disturbed during harvesting or other forestry practices should reasonably be included as an energy use?

Arvidsson et al. (2012) illustrated in the study on energy indicators for biofuels how different indicators represent different system boundaries. Depending on what aspects of the energy use that you should be held accountable for and what concerns that you want the indicator to represent, a careful selection should be made. If there is no point in punishing a product for an inefficient and fossil-based background energy system, for example because it could just as well have been produced in another part of the world with the same technology, accounting for energy use on a secondary energy level makes sense. Arvidsson and Svanström (2016) discuss the relevance of different energy use indicators to different actors.

It is thus clear that results vary greatly for some bio-based products when different energy use indicators are applied and it is therefore important for analysts to carefully select the most relevant indicator for individual studies of biobased products, and to be clear in the communication of their implications. Further, as one of the common differences between energy use indicators is whether they include renewable energy use or not, the discrepancies between different practices will be larger for product systems based on renewables, for example when forest biomass is increasingly employed. In fact, it can be argued that when the pressure increases on biologically productive systems, it would be wrong to include only fossil energy use as there are also limits on the availability of biomass. It can also be argued that biomass would possibly need to be separated from "flowing" renewables—e.g. solar radiation or air and water masses moving as a result of solar radiation—and accounted for in a category of its own.

4.3.6   Need for Improved Impact Assessment Terminology

There is a need to discuss impact assessment terminology, particularly in relation to the impacts referred to in the present book as "biodiversity loss" and "water cycle disturbances". In the LCA community, these impact categories are traditionally referred to with the less specific terms "land use" or "land use impact" and "water use" or "water use impact", respectively. The terminology in the present book was chosen to avoid the potential confusion surrounding the traditional terminology. This confusion arises mostly because of ambiguities regarding where water and land use impacts are placed on the cause-effect chain. Although water and land use impacts are often referred to and presented as mid-point impact categories (i.e. located in the middle of the cause-effect chain from stressor to environmental effects), the traditional proxies used to describe them are actually inventory-level indicators (i.e. stressors), while water and land use in themselves are activities in the product system (i.e. causes of the stressors). Furthermore, the cause-effect chains of water and land use impacts and other impact categories are interconnected and overlapping. For example, land use (and land use change) contributes to water cycle disturbances and climate change. Our suggestion to rename the mid-point impact category traditionally termed water use or water use impact to water cycle disturbances is an attempt to clarify its meaning and reduce confusion. Such specification of water-related impact categories has been proposed before, for example Quinteiro et al. (2015) suggested the terms "blue water availability due to land use" and "green water deprivation", and Heuvelmans et al. (2005) suggested the term "regional water balance". A similar solution for the mid-point impact category traditionally termed land use or land use impact is to be more precise in terms of the type of impact concerned, by referring to it as, for example, "physical habitat disturbances", "soil disturbances" or similar, depending on what the impact assessment method actually models. Further down the cause-effect chain, there is a further need to be careful with the terminology. For example, the impact category of "biodiversity loss" most often relates to the biodiversity loss caused by land use and land use change via, for example, physical habitat disturbances (as is also the case in the present book), although many other mid-point impact categories (e.g. eutrophication, acidification and climate change) also contribute to biodiversity loss (MA 2005). This confusion can be avoided by either specifying the intended type of biodiversity impact in more detail (e.g. "biodiversity loss due to physical habitat disturbances") or harmonising the end-point LCIA modelling of different mid-point impacts (e.g. so the impact category termed "biodiversity loss" in fact encompasses all the major drivers behind biodiversity loss). In any case, for effective and accurate communication of LCIA results, it is important to clearly distinguish between impact categories occurring at different places along the cause-effect chain (e.g. by presenting mid-point and end-point impact categories in separate figures) and to ensure that environmental impacts are not double-counted.

Considering the ongoing, extensive progress of more elaborate impact assessment methods for a wide range of environmental impacts, the LCA community must increasingly resolve confusions related to terminology and eventually build a consensus on the terminology surrounding emerging methods. This is an important task for consensus processes, such as the ones lead by the UNEP/SETAC life cycle initiative on land use impact assessment and WULCA on water use impact assessment. Clarity in terminology will improve the clarity of discussions aimed at managing conceptual and modelling challenges in LCA.

4.4 Managing Trade-offs and Connecting LCAs to Global Challenges

A common challenge in LCAs is how to make sense of LCIA results. This challenge can be broken down into questions such as: (i) What is small and what is big in terms of the product's environmental impact? (ii) How important are different impact categories? (iii) What is the potential of a product, or improvements of a product, for contributing to addressing regional or global environmental challenges? Traditional methods for weighting and normalisation can to some degree address these questions, as they can benchmark LCIA results in relation to regional- and global-scale impacts and policies (see Chap. 3). Questions (i) and (ii) above concern the management of potential trade-offs between impact categories; for example, in deciding between two technical options in designing a biorefinery, one alternative may result in lower climate impact but higher toxic impact, resulting in a trade-off between two impact categories. Multi-criteria decision analysis (MCDA) can be used for managing trade-offs; see Rowley et al. (2012) for a theoretical discussion of using MCDA in LCAs, and Cinelli et al. (2014) for a review of MCDA methods. Also, in R&D projects, team learning processes have been suggested as a means for managing trade-offs (Clancy et al. 2013). Regardless of method, it is commonly acknowledged that managing trade-offs between impact categories are inherently value-based activities (Finnveden et al. 2009; Hertwich and Hammitt 2001). This warrants LCA practitioners to be aware and transparent about such value-based activities that are made as part of the LCA work, which can clarify for decision-makers the consequences of different ways of interpreting the LCIA results.

The rapidly growing scientific understanding of the anthropogenic pressures on the Earth system and the risks of transgressing thresholds of biophysical processes—as manifested by the planetary boundaries framework (Steffen et al. 2015)—suggests that certain environmental impacts are, from a scientific viewpoint, more urgent to mitigate than others. This knowledge could potentially offer a new, scientifically more robust approach for managing trade-offs between impact categories. Indeed, there has been one attempt to derive a weighting method for LCAs from the planetary boundaries framework (Tuomisto et al. 2012). Also, recently the planetary boundaries framework was used as the basis in a procedure for quantifying absolute targets for impact reduction in LCA contexts (Sandin et al. 2015b). Apart from target setting, this procedure can be used for managing trade-offs and evaluate the potential of various interventions for reducing environmental impacts of products [i.e. addressing question (iii) above]. Understanding this can help those involved in improving, or developing new, forest products to evaluate whether potential environmental benefits represent a substantial contribution towards reducing environmental impacts, or merely a modest step that needs to be accompanied by other more drastic measures for impact reduction. However, Sandin et al. (2015b) showed that although the planetary boundaries framework can be used for interpreting LCIA results, the LCA practitioner still have to make value-based choices and assumptions when dividing the finite impact space allowed within the planetary boundaries to certain regions, market segments and products. In addition to the aforementioned attempts to use the planetary boundaries framework to aid in the interpretation of LCIA results, Jørgensen et al. (2014) attempted to use absolute limits for global environmental interventions (not referring to the planetary boundaries) in an LCIA method, and Bjørn et al. (2015) likewise recognised the potential of using the planetary boundaries framework in LCIA.

To summarise: recently there have been increasing research into how to operationalise the increasing knowledge about the global environmental challenges—for example as manifested in the planetary boundaries framework—into LCA. This is probably something we will see more of in the coming years, and it will facilitate linkages between LCA work and other environmental work among firms, sectors and governments (including strategic work, e.g. such as formulating visions and impact-reduction targets).

4.5 Integrating LCA Work in the R&D of New Products

A key challenge in the journey towards a bio-based economy is how to use LCA and other environmental assessment tools in the development of new forest products. There are numerous suggestions on how to integrate different methods for environmental assessment (often LCA) into R&D processes[3] (e.g. Chang et al. 2014; Clancy 2014; European Forest Institute 2014; Fazeni et al. 2014; Tambouratzis et al. 2014; Collado-Ruiz and Ostad-Ahmad-Ghorabi 2013; Askham et al. 2012; Devanathan et al. 2010; Manmek et al. 2010; Othman et al. 2010;

---

[3]The inclusion of environmental considerations in product development is sometimes referred to as "ecodesign", "sustainable product design", "design for the environment", "design for life cycle", "environmental product development" or similar. The definitions of, and the distinctions between, these terms are not further elaborated on in the present book.

Vinodh and Rathod 2010; Colodel et al. 2009; Kunnari et al. 2009; McAloone and Bey 2009; Ny 2009; Byggeth et al. 2007; Waage 2007; Rebitzer 2005; Nielsen and Wenzel 2002; Fleischer et al. 2001). These are often screening or simplified methods particularly designed for the assessment of preliminary product or process designs—see Rebitzer (2005) for a review of such methods.

What many of the ready-made methods and procedures have in common is their emphasis on a range of different sustainability criteria in addition to environmental ones (e.g. economic and social criteria) and the recognition of the need for some type of MCDA for handling potential trade-offs between different sustainability dimensions or impact categories. Also, many methods and procedures are primarily intended for assessments carried out in rather specific contexts—for instance in studies of certain product categories as noted by Hetherington et al. (2014)—sometimes with predefined impact assessment methodology adopted to that specific context.

Furthermore, the literature on environmental consideration in R&D most often focuses on *intra-organisational* R&D contexts (i.e. R&D carried out within a firm), which do not face the same degree of organisational complexity as the *inter-organisational* R&D contexts increasingly common in Europe. Inter-organisational R&D projects in Europe often involve firms, universities and research institutes from different countries and areas of expertise, with varying reasons for joining the project and various expectations of the project outcomes. This creates a high degree of organisational and cultural complexity, which can make it challenging to agree on the roles of LCA in the project, plan LCA work in accordance with the selected roles, influence decision-making in the project (e.g. in terms of the technical direction of the development work), and adapt the project work to unexpected LCA results. These challenges can make it difficult to utilise the full potential of LCA for assessing environmental impacts and influencing product development in an environmentally preferable direction. Because of these complexities, it is of utmost importance—already in the pre-project planning—to select appropriate roles for the LCA in the project. A further dive into the challenges of integrating LCA work in inter-organisational R&D projects, with some examples of solutions, is given in Sect. 4.5.1. Also, in inter-organisational contexts it may not be possible or even desirable to align and/or integrate the LCA work with the strategic long-term management of the product portfolio of a particular organisation as is often desirable in intra-organisational LCA work, and is thus a key element in many of the methods and procedures suggested in the aforementioned literature.

Although many solutions offered in literature on environmental consideration in R&D address case-specific challenges, they also address some recurring general challenges. Hetherington et al. (2014) draw on experiences from case studies in diverse sectors (nanotechnology, bioenergy and food) to identify four such general challenges: (i) comparability, for example the issue of comparing emerging technologies with existing commercial technologies in cases of incomparable functions and/or system boundaries; (ii) scaling issues, for example estimating the material and energy use of commercial scale production when the processes exist only at laboratory scale; (iii) data, for example issues of getting inventory data in time to be able to influence decision-making in the R&D process; and (iv) uncertainty, for example uncertain characterisation methods for the emissions of novel materials (e.g. regarding the toxicity of emissions of nanoparticles) or the inherent uncertainties of future product systems existing in a constantly changing world. These challenges are common for LCAs in general, but are particularly prominent for LCAs of future technologies.

4.5.1 Integrating LCA in Inter-organisational R&D Projects

Inter-organisational R&D projects often involve a mix of firms, universities and research institutes from different countries and disciplines. The participants often have different reasons for participating and diverse expectations of the project outcomes. Complexity is further enhanced as multiple activities are carried out in parallel to solve different technical problems. In this setting, it can be difficult to comprehend how different activities will interact and contribute towards the aim of the project. The projects are also often characterised by a focus on certain technical ideas or solutions—as it is often required to present a well-developed technical idea or solution to attract the funding—and once the funding has been secured, the application text (e.g. specifying tasks, milestones, deliverables and distribution of budget) and the competences of the project team can set limitations on what can be done in the project. Furthermore, although reduced environmental impact is often one of the stated driving forces of the project, the project work most often focusses on technical R&D, and therefore LCA work is allocated a relatively small share of the budget. LCA work may even be included not because of the wishes of the involved organisations but because of requirements from the commissioner or the funding agency. Some of the organisations and/or individuals involved in the project may therefore not see the value of, or not be particularly interested in, the LCA work and the LCA results. In such situations, LCA work may become an add-on that does not receive sufficient attention.

The above-described organisational complexities, including limited flexibility of the technical direction of the project and limited interest in the LCA work, can confine the possible roles of LCA and increase the importance of thorough and conscious selection of LCA roles. However, according to the experiences of the authors of the present book, the role selection is often done in an arbitrary manner. This leads to unclear LCA roles, which contributes further to differing and even contradicting expectations of the outcome of the LCA work among project participants, causing situations in which the full potential of LCA to assess environmental impacts and influence the technology development is not utilised. The challenge of role selection in LCAs in inter-organisational R&D projects is further described by Sandin et al. (2014a), which also suggest means for handling the challenges. First, the paper advocates a greater awareness about the many potential roles of LCA in R&D, and lists eight such roles: (i) guide the technical R&D, (ii) develop life-cycle thinking in the project team, (iii) support scale-up of the developed technology, (iv) direct future R&D activities, (v) market the developed technology, (vi) demonstrate inclusion of environmental concerns, (vii) contribute to LCA knowledge, and (viii) fulfil a requirement of the commissioner or funding agency that an LCA has to be performed in the project. Secondly, the paper suggests four characteristics of an R&D project that project commissioners, project managers and LCA practitioners can evaluate to improve the selection of the LCA role in the project: (i) the project's potential influence on environmental impacts of the developed technologies, (ii) the degrees of freedom available for the technical direction of the project, (iii) the project's potential to provide required input to the LCA, and (iv) access to relevant audiences for the LCA results. The paper recommends that these project characteristics are evaluated as early as possible in the planning of the project, as a basis for role selection, and that project planning then is done in a way that supports the selected roles. Important aspects of planning can be the timing of LCA activities in relation to other activities in the project (including delivery of data to the LCA work) and the communication plan for disseminating LCA results, with clearly defined audiences, internal and/or external to the project. After evaluation of the project characteristics, it may turn out that the desired roles are not available. In such situations, the project manager must remove barriers to the desirable roles, or convince those that request roles that the desired roles are unavailable, a process termed "expectation management" in management and organisation theory (Bosch-Sijtsema 2007). Finally, the paper gathered input also from non-European LCA practitioners, which suggested that the roles of LCA in R&D contexts can be confined not only by the project characteristics proposed above, but also by the traditions among funding agencies and LCA practitioners. This suggests that more wide-spread knowledge of the possible roles of LCA in R&D projects could expand the use of LCA in contexts where it is not traditionally used.

    To conclude, the research community has paid little attention to LCA applied in inter-organisational R&D projects—although some of the above-discussed findings probably represent tacit knowledge amongst LCA practitioners. Sandin et al. (2014a) thus opens up a new field of research. This was recently been followed by Baldassarri et al. (2016), who shared experiences from the integration of environmental aspects into a specific inter-organisational R&D project. Hopefully, more LCA practitioners will begin to share their experiences from similar contexts in the form of formalised research—this would be important for improving the use of LCA in inter-organisational R&D projects, which can be expected to be an increasingly common context for LCA work, at least in Europe. Improving the use of LCA in a certain context is a means for ensuring that the LCA has as much impact as possible in that context, which can be seen as improving the "intervention" capacity of the study, in the terminology by Sandén and Harvey (2008). Sandén and Harvey recognised that intervention activities are not often seen as an integral part of systems studies. They suggest that intervention activities warrant greater effort, and therefore suggest that intervention should be seen as one of four "core activities" of systems studies, along with problem selection, system design and assessment (Sandén and Harvey 2008). Increased focus on intervention activities, in the context of product development, has also been suggested by Clancy (2014).

References

Agostini A, Giuntilo J, Boulamanti A (2013) Carbon accounting of forest bioenergy: conclusions and recommendations from a critical literature review. JRC Technical Reports, Report EUR 25354 EN. http://iet.jrc.ec.europa.eu/bf-ca/sites/bf-ca/files/files/documents/eur25354en_online-final.pdf. Accessed Dec 2014

Ardente F, Beccali M, Cellura M, Mistretta M (2008) Building energy performance: a LCA case study of kenaf-fibres insulation board. Energy Build 40:1–10

Arvidsson R, Fransson K, Fröling M, Svanström M, Molander S (2012) Energy use indicators in energy and life cycle assessments of biofuels: review and recommendations. J Clean Prod 31:54–61

Arvidsson R, Baumann H, Hildenbrand J (2015a) On the scientific justification of the use of working hours, child labour and property rights in social life cycle assessment: three topical reviews. Int J Life Cycle Assess 20(2), 161–173

Arvidsson R, Ngyen D, Svanström M (2015b). Life cycle assessment of cellulose nanofibrils production by mechanical treatment and two different pretreatment processes. Environ Sci Technol 49(11), 6881–6890

Arvidsson R, Svanström M (2016) A framework for energy use indicators and reporting in life cycle assessment. Integr Environ Assess Manag. doi:10.1002/ieam.1735

Askham C, Gade AL, Hanssen OJ (2012) Combining REACH, environmental and economic performance indicators for strategic sustainable product development. J Clean Prod 35:71–78

Baldassarri C, Mathieux F, Ardente F, Wehmann C, Deese K (2016) Integration of environmental aspects into R&D inter-organizational projects management: application of a life cycle-based method to the development of innovative windows. J Clean Prod 112(4):3388–3401

Bayart J-B, Bulle C, Margni M, Vince F, Deschenes L, Aoustin E (2009) Operational characterisation method and factors for a new midpoint impact category: freshwater deprivation for human uses. In: Proceedings of the SETAC Europe 19th annual meeting, Gothenburg, Sweden

Bayart J-B, Worbe S, Grimaud J, Aoustin E (2014) The water impact index: a simplified single-indicator approach for water footprinting. Int J Life Cycle Assess 19:1336–1344

Berger M, Finkbeiner M (2010) Water footprinting: how to address water use in life cycle assessment? Sustainability 2(4):919–944

Berger M, Finkbeiner M (2012) Methodological challenges in volumetric and impact-oriented water footprints. J Ind Ecol 17(1):79–89

Bhattacharyya A, Mazumdar A, Roy PK, Sarkar A (2013) Life cycle assessment of carbon flow through harvested wood products. Ecol Environ Conserv 19(4):1195–1209

Bjørn A, Diamond M, Owsianiak M, Verzat B, Hauschild MZ (2015) Strengthening the link between life cycle assessment and indicators for absolute sustainability to support development within planetary boundaries. Environ Sci Technol 20(7):1005–1018

Blengini GA (2009) Life cycle of buildings, demolition and recycling potential: a case study in Turin, Italy. Build Environ 44:319–330

Börjeson L, Höjer M, Dreborg K-H, Ekvall T, Finnveden G (2005) Towards a user's guide to scenarios—a report on scenario types and scenario techniques. Environmental strategies research, Department of Urban studies, Royal Institute of Technology, Stockholm

Börjeson L, Höjer M, Dreborg K-H, Ekvall T, Finnveden G (2006) Scenario types and techniques: towards a user's guide. Futures 38(7):723–739

Bösch ME, Hellweg S, Huijbregts MAJ, Frischknecht R (2007) Applying cumulative exergy demand (CExD) indicators to the ecoinvent database. Int J Life Cycle Assess 12:181–190

Bosch-Sijtsema P (2007) The impact of individual expectations and expectation conflicts on virtual teams. Group Organ Manage 32(3):358–388

Bouhaya L, Le Roy R, Feraille-Fresnet A (2009) Simplified environmental study on innovative bridge structures. Environ Sci Technol 43:2066–2071

Boulay A-M, Motoshita M, Pfister S, Bulle C, Muñoz I, Franceschini H, Margni M (2015a) Analysis of water use impact assessment methods (part A): evaluation of modelling choices based on quantitative comparison of scarcity and human health indicators. Int J Life Cycle Assess 20(1):139–160

Boulay A-M, Bare J, De Camillis C, Doll P, Gassert F, Gerten D et al (2015b) Consensus building on the development of a stress-based indicator for LCA-based impact assessment of water consumption: outcome of the expert workshops. Int J Life Cycle Assess 20(5):577–583

Brandão M, Milà i Canals L (2013) Global characterisation factors to assess land use impacts on biotic production. Int J Life Cycle Assess 18(6):1243–1252

Brandão M, Milà i Canals L, Clift R (2011) Soil organic carbon changes in the cultivation of energy crops: implications for GHG balances and soil quality for use in LCA. Biomass Bioenergy 35(6):2323–2336

Brandão M, Levasseur A, Kirschbaum MUF, Weidema BP, Cowie AL, Vedel Jørgensen S et al (2013) Key issues and options in accounting for carbon sequestration and temporary storage in life cycle assessment and carbon footprinting. Int J Life Cycle Assess 18:230–240

Brander M (2015) Response to "Attributional life cycle assessment: is a land-use baseline necessary?"—appreciation, renouncement, and further discussion. Int J Life Cycle Assess. doi:10.1007/s11367-015-0974-8

Brentrup F, Küsters J, Lammel J, Kuhlmann H (2002) Life cycle impact assessment of land use based on the hemeroby concept. Int J Life Cycle Assess 7:339–348

Bribián IZ, Capilla AV, Usón AA (2011) Life cycle assessment of building materials: comparative analysis of energy and environmental impacts and evaluation of the eco-efficiency improvement potential. Build Environ 26:1133–1140

Bruijnzeel LA (2004) Hydrological functions of tropical forests: not seeing the soil for the trees? Agr Ecosyst Environ 104:185–228

Buyle M, Braet J, Audenaert A (2013) Life cycle assessment in the construction sector: a review. Renew Sustain Energy Rev 26:379–388

Byggeth S, Broman G, Robért K-H (2007) A method for sustainable product development based on a modular system of guiding questions. J Clean Prod 15:1–11

Cai Z, Laughlin R, Stevens R (2001) Nitrous oxide and dinitrogen emissions from soil under different water regimes and straw amendment. Chemosphere 42:113–121

Cellura M, Longo S, Mistretta M (2011) Sensitivity analysis to quantify uncertainty in life cycle assessment: the case study of an Italian tile. Renew Sustain Energy Rev 15:4697–4705

Chang D, Lee CKM, Chen C-H (2014) Review of life cycle assessment towards sustainable product development. J Clean Prod 83:48–60

Chapin FS III, Zavaleta ES, Eviner VT, Naylor RL, Vitousek PM, Reynolds HL et al (2000) Consequences of changing biodiversity. Nature 405:234–242

Cherubini F, Strømman AH, Ulgiati S (2011) Influence of allocation methods on the environmental performance of biorefinery products—a case study. Resour Conserv Recycl 55:1070–1077

Cherubini F, Bright RM, Strømman AH (2012) Site-specific global warming potentials of biogenic CO_2 for bioenergy: contributions from carbon fluxes and albedo dynamics. Environ Res Lett 7 (4). doi:10.1088/1748-9326/7/4/045902

Cherubini F, Guest G, Strømman AH (2013) Bioenergy from forestry and changes in atmospheric CO_2: reconciling single stand and landscape level approaches. J Environ Manage 129:292–301

Cinelli M, Coles SR, Kirwan K (2014) Analysis of the potentials of multi criteria decision analysis methods to conduct sustainability assessment. Ecol Ind 46:138–148

Clancy G (2014) Assessing sustainability and guiding development towards more sustainable products. Thesis for the degree of doctor of philosophy, Chalmers University of Technology, Chalmers Reproservice, Gothenburg, Sweden. http://publications.lib.chalmers.se/publication/197988. Accessed Nov 2014

Clancy G, Fröling M, Svanström M (2013) Insights from guiding material development towards more sustainable products. Int J Sustain Des 2(2):149–166

Collado-Ruiz D, Ostad-Ahmad-Ghorabi H (2013) Estimating environmental behaviour without performing a life cycle assessment. J Ind Ecol 17(1):31–42

Colodel MC, Kupfer T, Barthel L-P, Albrecht S (2009) R&D decision support by parallel assessment of economic, ecological and social impact—adipic acid from renewable resources versus adipic acid from crude oil. Ecol Econ 68(6):1599–1604

Costa PM, Wilson C (2000) An equivalence factor between CO_2 avoided emissions and sequestration—description and applications in forestry. Mitig Adapt Strat Glob Change 5:51–60

Cuéllar-Franca RM, Azapagic A (2012) Environmental impacts of the UK residential sector: life cycle assessment of houses. Build Environ 54:86–99

Curran M, de Baan L, de Schryver A, van Zelm R, Hellweg S, Koellner S et al (2011) Toward meaningful end points of biodiversity in life cycle assessment. Environ Sci Technol 45:70–79

de Baan L, Alkemede R, Koellner T (2012) Land use impacts on biodiversity in LCA: a global approach. Int J Life Cycle Assess 18(6):1216–1230

de Baan L, Mutel CL, Curran M, Hellweg S, Koellner T (2013) Land use in life cycle assessment: global characterization factors based on regional and global potential species extinction. Environ Sci Technol 47:9281–9290

De Souza DM, Flynn DFB, DeClerck F, Rosenbaum RK, de Melo Lisboa H, Koellner T (2013) Land use impacts on biodiversity in LCA: proposal of characterization factors based on functional diversity. Int J Life Cycle Assess 18(6):1231–1242

De Souza DM, Teixeira RFM, Ostermann OP (2015) Assessing biodiversity loss due to land use with life cycle assessment: are we there yet? Glob Change Biol 21:32–47

Devanathan S, Ramanujan D, Bernstein WZ, Zhao F, Ramani K (2010) Integration of sustainability into early design through the function impact matrix. J Mech Des 132

Dixit MK, Fernández-Solís JL, Lavy S, Culp CH (2012) Need for an embodied energy measurement protocol for buildings: a review paper. Renew Sustain Energy Rev 16:3730–3743

Dornburg V, Marland G (2008) Temporary storage of carbon in the biosphere does have value for climate change mitigation: a response to the paper by Miko Kirschbaum. Mitig Adapt Strat Glob Change 13(3):211–217

Du G, Mohammed S, Pettersson L, Karoumi R (2014) Life cycle assessment as a decision support tool for bridge procurement: environmental impact comparison among five bridge designs. Int J Life Cycle Assess 19:1948–1968

EC (2009a) Directive 2009/30/EC of the European Parliament and of the Council of 23 April 2009. http://eur-lex.europa.eu/LexUriServ/LexUriServ.do?uri=OJ:L:2009:140:0088:0113:EN:PDF. Accessed Jan 2015

EC (2009b) Directive 2009/28/EC of the European Parliament and of the Council of 23 April 2009. http://eur-lex.europa.eu/legal-content/EN/TXT/PDF/?uri=CELEX:32009L0028&from=en. Accessed Jan 2015

EC (2010) International reference life cycle data system (ILCD) handbook—general guide for the life cycle assessment—detailed guidance. Joint Research Centre—Institute for Environment and Sustainability. Publications Office of the European Union, Luxembourg

EC (2013) Commission recommendation of 9 April 2013 on the use of common methods to measure and communicate the life cycle environmental performance of products and organisations. http://eur-lex.europa.eu/legal-content/EN/TXT/PDF/?uri=CELEX:32013H0179&from=EN. Accessed Feb 2015

European Environment Agency (1995) CORINE land cover. European Environment Agency. http://www.eea.europa.eu/publications/COR0-landcover. Accessed Jan 2015

European Forest Institute (2014) ToSIA—tool for sustainable impact assessment. http://tosia.efi.int/. Accessed Jan 2015

European Space Agency (2011) GlobCover. http://due.esrin.esa.int/globcover. Accessed Jan 2015

Fazeni K, Lindorfer J, Prammer H (2014) Methodological advancements in life cycle process design: a preliminary outlook. Resour Conserv Recycl 92:66–77

Finnveden G, Hauschild MZ, Ekvall T, Guinée J, Heijungs R, Hellweg S et al (2009) Recent developments in life cycle assessment. J Environ Manage 91:1–21

Fisher B, Turner RK, Morling P (2009) Defining and classifying ecosystem services for decision making. Ecol Econ 68:643–653

Fleischer G, Gerner K, Kunst H, Licthenvort K, Rebitzer G (2001) A semi-quantitative method for the impact assessment of emissions within a simplified life cycle assessment. Int J Life Cycle Assess 6(3):149–156

Frijia S, Guhathakurta S, Williams E (2011) Functional unit, technological dynamics, and scaling properties for the life cycle of residencies. Environ Sci Technol 46:1782–1788

Frischknecht R, Büsser S, Krewitt W (2009a) Environmental assessment of future technologies: how to trim LCA to fit this goal. Int J Life Cycle Assess 14:584–588

Frischknecht R, Steiner R, Jungbluth N (2009b) The ecological scarcity method—eco-factors 2006. A method for impact assessment in LCA. Environmental studies no. 0906. Federal Office for the Environment, Bern

FSC (2016) https://ic.fsc.org. Accessed Jan 2016

Gamfeldt L, Snäll T, Bagchi R, Jonsson M, Gustafsson L, Kjellander P et al (2013) Higher levels of multiple ecosystem services are found in forests with more tree species. Nat Commun 4. doi:10.1038/ncomms2328

Garcia R, Freire A (2014) Carbon footprint of particleboard: a comparison between ISO/TS 14067, GHG protocol, PAS 2050 and climate declaration. J Clean Prod 66:199–200

Geyer R, Lindner J, Stoms D, Davis F, Wittstock B (2010) Coupling GIS and LCA for biodiversity assessments of land use, part 2: impact assessment. Int J Life Cycle Assess 15:692–703

Goedkoop M, Spriensma R (2000) The eco-indicator 99—a damage-oriented method for life cycle impact assessment, 2nd edn. PRé Consultants, Amersfoort

Grant A, Ries R, Kibert C (2014) Life cycle assessment and service life prediction: a case study of building envelope materials. J Ind Ecol 18(2):187–200

Guest G, Strømman AH (2014) Climate change impacts due to biogenic carbon: addressing the issue of attribution using two metrics with very different outcomes. J Sustain Forest 33 (3):298–326

Guest G, Cherubini F, Strømman AH (2013) Global warming potential of carbon dioxide emissions from biomass stored in the anthroposphere and used for bioenergy at end of life. J Ind Ecol 17(1):20–30

Guinée JB, Gorrée M, Heijungs R, Huppes G, Kleijn R, Koning A (2002) Handbook on life cycle assessment. Operational guide to the ISO standards. Kluwer Academic Publishers, Dordrecht

Gunn JS, Ganz D, Keeton W (2012) Biogenic vs. geologic carbon emissions and forest biomass energy production. GCB Bioenergy 4:239–242

Habert G, Arribe D, Dehove T, Espinasse L, Le Roy R (2012) Reducing environmental impact by increasing the strength of concrete: quantification of the improvement to concrete bridges. J Clean Prod 35:250–262

Haines-Young R, Potschin M (2013) Common international classification of ecosystem services (CICES): consultation on version 4, August–December 2013. EEA Framework Contract No EEA/IEA/09/003. http://cices.eu/wp-content/uploads/2012/07/CICES-V43_Revised-Final_Report_29012013.pdf. Accessed Dec2014

Heijungs R, Guinée JB (2007) Allocation and 'what-if' scenarios in life cycle assessment of waste management systems. Waste Manage 27:997–1005

Heijungs R, Huppes G, Guinée J (2009) A scientific framework for LCA. Deliverable (D15) of work package 2 (WP2) CALCAS project. http://www.leidenuniv.nl/cml/ssp/publications/calcas_report_d15.pdf. Accessed Dec 2014

Helin T, Sokka L, Soimakallio S, Pingoud K, Pajula T (2013) Approaches for inclusion of carbon cycles in life cycle assessment—a review. GCB Bioenergy 5(5):475–486

Hertwich EG, Hammitt JK (2001) A decision-analytic framework for impact assessment. Part 2: midpoints, endpoints, and criteria for method development. Int J Life Cycle Assess 6(1):5–12

Hetherington AC, Borrion AL, Griffiths OG, McManus MC (2014) Use of LCA as a development tool within early research: challenges and issues across different sectors. Int J Life Cycle Assess 19:130–143

Heuvelmans G, Muys G, Feyen J (2005) Extending the life cycle methodology to cover impacts of land use systems on the freshwater balance. Int J Life Cycle Assess 10:113–119

Hoekstra AY, Chapagain AK (2007) Water footprints of nations: water use by people as a function of their consumption pattern. Water Resour Manage 21(1):35–48

Hoekstra AY, Chapagain AK, Aldaya MM, Mekonnen MM (2011) The water footprint assessment manual: setting the global standard. Water Footprint Network, Enschede

Hooper DU, Adair EC, Cardinale BR, Byrnes JEK, Hungate BA, Matulich KL et al (2012) A global synthesis reveals biodiversity loss as a major driver of ecosystem change. Nature 486:105–108

Huijbregts MAJ, Hellweg S, Frischknecht K, Hendriks HWM, Hungerbühler K, Hendriks AJ (2010) Cumulative energy demand as predictor for the environmental burden of commodity production. Environ Sci Technol 44(6):2189–2196

IPCC (2013) In: Stocker TF, Qin D, Plattner G-K, Tignor M, Allen SK, Boschung J et al (eds) Climate change 2013: the physical science basis. Working group I contribution to the 5th assessment report of the Intergovernmental Panel on Climate Change. Cambridge University Press, Cambridge, UK and New York, NY, USA. http://www.ipcc.ch/report/ar5/wg1/. Accessed Oct 2014

ISO (2006a) 14040: Environmental management—life cycle assessment—requirements and guidelines. International Organisation for Standardisation

ISO (2006b) 14044: Environmental management—life cycle assessment – principles and framework. International Organisation for Standardisation

ISO (2014) 14046: Environmental management—water footprint—principles, requirements and guidelines. International Organisation for Standardisation

Johnson E (2009) Goodbye to carbon neutral: getting biomass footprints right. Environ Impact Assess Rev 29:165–168

Jørgensen SV, Hauschild MZ, Nielsen PH (2014) Assessment of urgent impacts of greenhouse gas emissions—the climate tipping potential (CTP). Int J Life Cycle Assess 19(4):919–930

Karlsson H, Börjesson P, Hansson P-H, Ahlgren S (2014) Ethanol production in biorefineries using lignocellulosic feedstock—GHG performance, energy balance and implications of life cycle calculation methodology. J Clean Prod 83:420–427

Kirschbaum MUF (2006) Temporary carbon sequestration cannot prevent climate change. Mitig Adapt Strat Glob Change 11(5–6):1151–1164

Klein D, Wolf C, Schulz C, Blaschke-Weber G (2015) 20 years of life cycle assessment (LCA) in the forestry sector: state of the art and a methodological proposal for the LCA of forest production. Int J Life Cycle Assess 20:556–575

Koellner T (2000) Species-pool effect potentials (SPEP) as a yardstick to evaluate land-use impacts on biodiversity. J Clean Prod 8:293–311

Koellner T, Geyer R (2013) Global land use impact assessment on biodiversity. Int J Life Cycle Assess 18(6):1185–1187

Koellner T, Scholz RW (2008) Assessment of land use impacts on the natural environmental. Int J Life Cycle Assess 13(1):32–48

Koellner T, de Baan L, Beck T, Brandão M, Civit B, Goedkoop M et al (2013a) Principles for life cycle inventories of land use on a global scale. Int J Life Cycle Assess 18(6):1203–1215

Koellner T, de Baan L, Beck T, Brandão M, Civit B, Goedkoop M et al (2013b) UNEP-SETAC guideline on global land use impact assessment on biodiversity and ecosystem services in LCA. Int J Life Cycle Assess 18(6):1188–1202

Koponen K, Soimakallio S (2015) Foregone carbon sequestration due to land occupation—the case of agro-bioenergy in Finland. Int J Life Cycle Assess 20(11):1544–1556

Kounina A, Margni M, Bayart J-B, Boulay A-M, Berger M, Bulle C et al (2013) Review of methods addressing freshwater use in life cycle inventory and impact assessment. Int J Life Cycle Assess 18(3):701–721

Kunnari E, Valkama J, Keskinen M, Mansikkamäki P (2009) Environmental evaluation of new technology: printed electronics case study. J Clean Prod 17:791–799

Kyläkorpi L, Rydgren B, Ellegård A, Miliander S, Grusell E (2005) The biotope method 2005: a method to assess the impact of land use on biodiversity. http://www.vattenfall.com/en/file/2005TheBiotopeMethod_8459811.pdf. Accessed Jan 2013

Launiainen S, Futter MN, Ellison D, Clarke N, Finér L, Högbom L et al (2014) Is the water footprint an appropriate tool for forestry and forest products: the Fennoscandian case. Ambio 43(2):244–256

Levasseur A, Lesage P, Margni M, Deschênes L, Samson R (2010) Considering time in LCA: dynamic LCA and its application to global warming impact assessments. Environ Sci Technol 44:3169–3174

Lindeijer E (2000) Biodiversity and life support impacts of land use in LCA. J Clean Prod 8:313–319

Lindner JP, Niblick B, Eberle U, Bos U, Schmincke E, Schwarz S, et al. (2014) Proposal of a unified biodiversity impact assessment method. In: Proceedings of the 9th international conference LCA of food, San Francisco, USA

Lindqvist M, Palme U, Lindner JP (2015) A comparison of two different biodiversity assessment methods in LCA—a case study of Swedish spruce forest. Int J Life Cycle Assess. doi:10.1007/s11367-015-1012-6

Liski J, Korotkov AV, Prins CFL, Karjalainen T, Victor DG, Kauppi PE (2003) Increased carbon sink in temporal and boreal forests. Clim Change 61:89–99

Lundie S, Peters G, Beavis P (2004) Life cycle assessment for sustainable metropolitan water systems planning—options for ecological sustainability. Environ Sci Technol 38:3465–3473

Luo L, Van Der Voet E, Huppes G, Udo De Haes HA (2009) Allocation issues in LCA methodology: a case study of corn stover-based fuel ethanol. Int J Life Cycle Assess 14(6):529–539

MA (2005) Ecosystems and human well-being: biodiversity synthesis. World Resources Institute, Washington DC

Manmek S, Kaebernick H, Kara S (2010) Simplified environmental impact drivers for product life cycle. Int J Sustain Manuf 2(1):30–65

Mathiesen BV, Münster M, Fruergaard T (2009) Uncertainties related to the identification of the marginal technology in consequential life cycle assessments. J Clean Prod 17:1331–1338

Mattsson B, Cederberg C, Blix L (2000) Agricultural land use in life cycle assessment (LCA): case studies of three vegetable oil crops. J Clean Prod 8:283–292

McAloone TC, Bey N (2009) Environmental improvement through product development: a guide. Danish Environmental Protection Agency, Copenhagen

Michelsen O (2008) Assessment of land use impact on biodiversity: proposal of a new methodology exemplified with forestry operations in Norway. Int J Life Cycle Assess 13(1):22–31

Michelsen O, Cherubini F, Strømman AH (2012) Impact assessment of biodiversity and carbon pools from land use and land use change in life cycle assessment, exemplified with forestry operations in Norway. J Ind Ecol 16(2):231–242

Milà i Canals L, Romanyà J, Cowell SJ (2007) Method for assessing impacts on life support functions (LSF) related to the use of 'fertile land' in life cycle assessment (LCA). J Clean Prod 15:1426–1440

Milà i Canals L, Chenoweth J, Chapagain A, Orr S, Antón A, Clift R (2009) Assessing freshwater use in LCA: part I—inventory modelling and characterisation factors for the main impact pathways. Int J Life Cycle Assess 14:28–42

Motoshita M, Itsubo N, Inaba A (2008). Development of impact assessment method on health damages of undernourishment related to agricultural water scarcity. In: Proceedings of the 8th international conference on EcoBalance, Tokyo, Japan

Motoshita M, Itsubo N, Inaba A, Aoustin E (2009) Development of damage assessment model for infectious diseases arising from domestic water consumption. In: Proceedings of the SETAC Europe: 19th annual meeting, Gothenburg, Sweden

Muys B, Quijano JG (2002) A new method for land use impact assessment in LCA based on the ecosystem exergy concept. http://www.biw.kuleuven.be/lbh/lbnl/forecoman/pdf/land%20use%20method4.pdf. Accessed Mar 2015

Nielsen PH, Wenzel H (2002) Integration of environmental aspects in product development: a stepwise procedure based on quantitative life cycle assessment. J Clean Prod 10:247–257

Núñez M, Antón A, Muñoz P, Rieradevall J (2013) Inclusion of soil erosion impacts in life cycle assessment on a global scale: application to energy crops in Spain. Int J Life Cycle Assess 18:755–767

Ny H (2009) Strategic life-cycle modeling and simulation for sustainable product development. Blekinge Institute of Technology Doctoral Dissertation Series No. 2009:02. http://www.bth.se/fou/forskinfo.nsf/all/d218ba0b67bf3802c12575b400295b6b/$file/Ny_diss.pdf. Accessed Feb 2015

Olson DM, Dinerstein E, Wikramanayake ED, Burgess ND, Powell GVN, Underwood EC et al (2001) Terrestrial ecoregions of the world: a new map of life on Earth. Bioscience 51 (11):933–938

Ortiz O, Pasqualino JC, Castells F (2010) The environmental impact of the construction phase: an application to composite walls from a life cycle perspective. Resour Conserv Recycl 54:832–840

Othman MR, Repke J-U, Wozny G, Huang Y (2010) A modular approach to sustainability assessment and decision support in chemical process design. Ind Eng Chem Res 49:7870–7881

Pawelzik P, Carus M, Hotchkiss J, Narayan R, Selke S, Wellisch M et al (2013) Critical aspects in the life cycle assessment (LCA) of bio-based materials—reviewing methodologies and deriving recommendations. Resour Conserv Recycl 73:211–228

PEFC (2016) http://www.pefc.org. Accessed January 2016

Pelletier N, Ardente F, Brandão M, De Camillis C, Pennington D (2015) Rationales for and limitations of preferred solutions for multi-functionality problems in LCA: is increased consistency possible? Int J Life Cycle Assess 20(1):74–86

Perez-Garcia J, Lippke B, Comnick J, Manriquez C (2005) An assessment of carbon pools, storage, and wood products market substitution using life-cycle analysis results. Wood Fiber Sci 37:140–148

Persson C, Fröling M, Svanström M (2006) Life cycle assessment of the district heat distribution system, part 3: use phase and overall discussion. Int J Life Cycle Assess 11:437–446

Pesonen H-L, Ekvall T, Fleischer G, Huppes G, Jahn C, Klos SZ et al (2000) Framework for scenario development in LCA. Int J Life Cycle Assess 5:21–30

Peters GM, Wiedemann SG, Rowley HV, Tucker RV (2010) Accounting for water use in Australian red meat production. Int J Life Cycle Assess 15(3):311–320

Peters GM, Blackburn NJ, Armedio M (2013) Environmental assessment of air to water machines—triangulation to manage scope uncertainty. Int J Life Cycle Assess 18:1149–1157

Pfister S, Koehler A, Hellweg S (2009) Assessing the environmental impacts of freshwater consumption in LCA. Environ Sci Technol 43:4098–4104

Pinsonnault A, Lesage P, Levasseur A, Samson R (2014) Temporal differentiation of background systems in LCA: relevance of adding temporal information in LCI databases. Int J Life Cycle Assess 19:1843–1853

Quinteiro P, Cláudia Dias A, Silva M, Ridoutt BG, Arroja L (2015) A contribution to the environmental impact assessment of green water flows. J Clean Prod. doi:10.1016/j.jclepro. 2015.01.022

Reap J, Roman F, Duncan S, Bras B (2008) A survey of unresolved problems in life cycle assessment, part 2: impact assessment and interpretation. Int J Life Cycle Assess 13:374–388

Rebitzer G (2005) Enhancing the application efficiency of life cycle assessment for industrial uses. Thesis no 3307, École polytechnique fédérale de Lausanne, Switzerland. http://infoscience. epfl.ch/record/52216/files/EPFL_TH3307.pdf. Accessed Jan 2015

Repo A, Tuomi M, Liski J (2011) Indirect carbon dioxide emissions from producing bioenergy from forest harvest residues. GCB Bioenergy 3(2):107–115

Ridoutt BG (2011) Development and application of water footprint metric for agricultural products and the food industry. In: Finkbeiner M (ed) Towards life cycle sustainability management. Springer, Dordrecht, pp 183–192

Ridoutt BG, Pfister S (2013) A new water footprint calculation method integrating consumptive and degradative water use into a single stand-alone weighted indicator. Int J Life Cycle Assess 18:204–207

Ridoutt BG, Sanguansri P, Nolan M, Marks N (2012) Meat consumption and water scarcity: beware of generalizations. J Clean Prod 28:127–133

Rowley HV, Peters GM, Lundie S, Moore SJ (2012) Aggregating sustainability indicators: beyond the weighted sum. J Environ Manage 111:24–33

Røyne F, Peñaloza D, Sandin G, Berlin J, Svanström M (2016) Climate impact assessment in life cycle assessments of forest products: implications of method choice for results and decision-making. J Clean Prod 116:90–99

Saad R, Margni M, Koellner T, Wittstock B, Deschênes L (2011) Assessment of land use impacts on soil ecological functions: development of spatially differentiated characterization factors within a Canadian context. Int J Life Cycle Assess 16:198–211

Sandén BA, Harvey S (2008) System analysis for energy transition: a mapping of methodologies, co-operation and critical issues in energy systems studies at Chalmers. Report CEC 2008:2, Chalmers University of Technology, Gothenburg, Sweden

Sandin G, Peters GM, Svanström M (2013) Moving down the cause-effect chain of water and land use impacts: an LCA case study of textile fibres. Resour Conserv Recycl 17:104–113

Sandin G, Peters GM, Svanström M (2014a) Life cycle assessment of construction materials: the influence of assumptions in end-of-life modelling. Int J Life Cycle Assess 19(4):723–731

Sandin G, Clancy G, Heimersson S, Peters GM, Svanström M, ten Hoeve M (2014b) Making the most of LCA in inter-organisational R&D projects. J Clean Prod 70:97–104

Sandin G, Røyne F, Berlin H, Peters GM, Svanström M (2015a) Allocation in LCAs of biorefinery products: implications for results and decision-making. J Clean Prod 93:213–221

Sandin G, Peñaloza D, Røyne F, Svanström M, Staffas L (2015b) The method's influence on climate impact assessment of biofuels and other uses of forest biomass. Report No 2015:10, f3 The Swedish Knowledge Centre for Renewable Transportation Fuels, Sweden. www.f3centre.se. Accessed Jan 2016

Sandin G, Peters GM, Svanström M (2015c) Using the planetary boundaries framework for setting impact-reduction targets in LCA contexts. Int J Life Cycle Assess 20(12):1684–1700

Schmidt JH (2008) Development of LCIA characterisation factors for land use impacts on biodiversity. J Clean Prod 16:1929–1942

Schmidt JH, Weidema BP, Brandão M (2015) A framework for modelling indirect land use changes in life cycle assessments. J Clean Prod. doi:10.1016/j.jclepro.2015.03.013

Schulze E-D, Körner C, Law BE, Haberl H, Luyssaert S (2012) Large-scale bioenergy from additional harvest of forest biomass is neither sustainable nor greenhouse gas neutral. GCB Bioenergy 4:611–616

Schwaiger H, Bird N (2010) Integration of albedo effects caused by land use change into the climate balance: should we still account in greenhouse gas units? Forest Ecol Manage 260:278–286

Searchinger T, Heimlich R, Houghton RA, Dong F, Elobeid A, Fabiosa J et al (2008) Use of U.S. croplands for biofuels increases greenhouse gases through emission from land-use change. Science 319:1238–1240

Singh A, Berghorn G, Joshi S, Syal M (2011) Review of life-cycle assessment applications in building construction. J Arch Eng 17:15–23

Sjølie HK, Solberg B (2011) Greenhouse gas emission impacts of use of Norwegian wood pellets: a sensitivity analysis. Environ Sci Policy 14(8):1028–1040

Soimakallio S, Cowie A, Brandão M, Finnveden G, Ekvall T, Erlandsson M et al (2015) Attributional life cycle assessment: is a land-use baseline necessary. Int J Life Cycle Assess 20 (10):1364–1375

Spielmann M, Scholz RW, Tietje O, de Haan P (2005) Scenario modelling in prospective LCA of transport systems: application of formative scenario analysis. Int J Life Cycle Assess 10 (5):325–335

Spracklen DV, Bonn B, Carslaw KS (2008) Boreal forests, aerosols and the impacts on clouds and climate. Philos Trans R Soc Lond A: Math Phys Eng Sci 366(1885):4613–4626

Steffen W, Richardson K, Rockström J, Cornell SE, Fetzer I, Bennett EM et al (2015) Planetary boundaries: guiding human development on a changing planet. Science. doi:10.1126/science.1259855

Stephenson AL, Dupree P, Scott SA, Dennis JS (2010) The environmental and economic sustainability of potential bioethanol from willow in the UK. Bioresour Technol 101(24):9612–9623

Sterman JD (1991) A skeptic's guide to computer models. In: Barney GO, Kreutzer WB, Garrett MJ (eds) Managing a nation: the microcomputer software catalog, 2nd edn. Westview Press, Boulder

Swank WT, Vose JM, Elliot KJ (2001) Long-term hydrologic and water quality responses following commercial clearcutting of mixed hardwoods on a southern Appalachian catchment. Forest Ecol Manage 143:163–178

Swedish University of Agricultural Sciences (2011). Forestry statistics 2011. http://www.slu.se/Global/externwebben/nl-fak/mark-och-miljo/Markinventeringen/Dokument%20MI/Skogsdata2011_temadelen%20om%20markvegetation.pdf. Accessed Jan 2015

Tambouratzis T, Karalekas D, Moustakas N (2014) A methodological study for optimizing material selection in sustainable product design. J Ind Ecol 18(4):508–516

Teixeira RFM, de Souza DM, Curran MP, Antón A, Michelsen O, Milà i Canals L (2015) Towards consensus on land use impacts on biodiversity in LCA: UNEP/SETAC Life Cycle Initiative preliminary recommendations based on expert contribution. J Clean Prod. doi:10.1016/j.jclepro.2015.07.118

Ter-Mikaelian MT, Colombo SJ, Chen J (2015) The burning question: does forest bioenergy reduce carbon emissions? A review of common misconceptions about forest carbon accounting. J. Forest 113(1):57–68

Thompson I (2011) Biodiversity, ecosystem thresholds, resilience and forest degradation. Unasylva 238(62). http://www.fao.org/docrep/015/i2560e/i2560e05.pdf. Accessed Dec 2014

Thormark C (2002) A low energy building in a life cycle—its embodied energy, energy need for operation and recycling potential. Build Environ 37:429–435

Tillman AM (2000) Significance of decision-making for LCA methodology. Environ Impact Assess Rev 20(1):113–123

Tuomisto HI, Hodge IH, Riordan P, Macdonald DW (2012) Exploring a safe operating approach to weighting in life cycle impact assessment—a case study of organic, conventional and integrated farming systems. J Clean Prod 37:147–153

Van Zelm R, Rombouts M, Snepvangers J, Huijbregts MAJ, Aoustin E (2009). Characterization factors for groundwater extraction based on plant species occurrence in the Netherlands. In: Proceedings of the SETAC Europe, 19th annual meeting, Gothenburg, Sweden

Verbeeck G, Hens H (2007) Life cycle optimization of extremely low energy dwellings, J Build Phys 31(2):143–178

Vinodh S, Rathod G (2010) Integration of ECQFD and LCA for sustainable product design. J Clean Prod 18(8):832–844

Violle C, Navas M-L, Vile D, Kazakou E, Fortunel C, Hummel I et al (2007) Let the concept of trait be functional! Oikos 116:882–892

Vogtländer J, Van Der Velden N, Van Der Lugt P (2014) Carbon sequestration in LCA, a proposal for a new approach based on the global carbon cycle; cases on wood and on bamboo. Int J Life Cycle Assess 19(1):13–23

Waage SA (2007) Re-considering product design: a practical "road-map" for integration of sustainability issues. J Clean Prod 15:638–649

Weidema B (2014) Has ISO 14040/44 failed its role as a standard for life cycle assessment? J Ind Ecol 18:324–326

Weidema BP, Lindeijer E (2001) Physical impacts of land use in product life cycle assessment. Final report of the EURENVIRON 1296 LCAGAPS sub-project on land use. Department of Manufacturing Engineering and Management, Technical University of Denmark, Lyngby, Denmark

Wrage N, van Groeningen JW, Oenema O, Baggs EM (2005) A novel dual-isotope labelling method for distinguishing between soil sources of N_2O. Rapid Commun Mass Spectrom 19:3298–3306

WULCA (2014) Consensual indicator project. http://www.wulca-waterlca.org/project.html. Accessed Jan 2016

Zamagni A, Guinée J, Heijungs R, Masoni P, Raggi A (2012) Lights and shadows in consequential LCA. Int J Life Cycle Assess 17:904–918

Zanchi G, Pena N, Bird N (2012) Is woody bioenergy carbon neutral? A comparative assessment of emissions from consumption of woody bioenergy and fossil fuel. GCB Bioenergy 4 (6):761–772

Chapter 5
Future Research Needs

Abstract This chapter summarises future research needs related to improving the methods and practices of life cycle assessment (LCA) of forest products. Among others, the research needs concern scenario modelling, attributional and consequential modelling, end-of-life modelling, impact assessment (of climate change, biodiversity loss, and water cycle disturbances), reducing confusion surrounding terminology, and integrating LCA in R&D settings.

Keywords Methodology · Context alignment · Inventory analysis · Characterisation method · Water use · Land use · Carbon footprint

This chapter summarises future research needs related to the content of this book. Hopefully, this can function as inspiration for readers interested in contributing to LCA research.

Regarding improved modelling of product systems, there is a need for further research on scenario modelling in LCA, for example on how to validate the usefulness of generated scenarios, or on what types of scenario modelling are suitable in different R&D contexts. As a suggestion, such research could review early attempts of scenario modelling published in the literature and compare the generated scenarios with the actual outcomes. For example, by studying LCAs carried out in the past (e.g. 20 years ago), it would be possible to study how well different approaches for end-of-life modelling have managed to capture actual end-of-life practices.

Furthermore, there is a need for more research on attributional and consequential modelling in LCA, for example in relation to the modelling of end-of-life processes of long-lived products or the handling of multi-functionality. It is desirable to build a consensus within the LCA community on the circumstances under which attributional or consequential approaches (or hybrid approaches including elements of both), as well as specific methods for handling multi-functionality, are preferable. This would help LCA practitioners make conscious and context-aligned choices of modelling approaches, which are not typical today (Zamagni et al. 2012). To facilitate this, there is a need to further evaluate the consequences of different methodological choices in different contexts, for example by using ready-made

G. Sandin et al., *Life Cycle Assessment of Forest Products*,
Biobased Polymers, DOI 10.1007/978-3-319-44027-9_5

decision-context classifications, such as those provided by the ILCD handbook (EC 2010). There is also a need for clarifying existing guidance on this matter, for example the guidance provided by the ILCD handbook (Ekvall et al. 2016).

There is a need for more case studies applying emerging LCIA methods for climate change, biodiversity loss, water cycle disturbances and other location-dependent impacts of forestry and forest products. In particular, there is a need for case studies that compare different methods with regard to their applicability in various contexts and how well their outcomes reflect realities, for example as done by Røyne et al. (2016) for climate impact assessment. Such research can contribute to concrete recommendations for increasingly context-aligned practices, for example in standards and guidelines. Also, there is a need for research that addresses specific issues related to LCIA methods. Concerning the consequential LCI approach for quantifying water cycle alterations proposed by Sandin et al. (2013), there is a need to test its applicability in further case studies and explore its potential advantages in further detail, such as the capturing of environmental disturbances due to increased runoff. Concerning the assessment of biodiversity loss, there is a need to use case studies to compare different species richness indicators and how well they reflect realities, and explore how to combine such indicators with other indicators for biodiversity or ecosystem quality. Good examples in this direction are the studies by de Baan et al. (2012), on the correlation between an indicator on relative species richness and other biodiversity indicators, and Lindqvist et al. (2015), comparing the applicability of two LCIA methods based on species richness and ecosystem indicators, respectively, in a specific context. However, to facilitate such comparisons, many of the most promising methods proposed in the literature need to be made less dependent on data that are only available for specific regions or product categories, either by adapting the methods to suit available data or by collecting data for a wider range of regions and product categories. Also, there is a need to further discuss how to allocate transformational impacts of forestry and agriculture between the first harvest after transformation and subsequent harvests, and eventually reach a consensus on a standard amortisation period, for example 20 years, as proposed by Koellner et al. (2013). Further examples of research needs related to biodiversity impact assessment were recently identified by De Souza et al. (2015).

The LCA community should work towards resolving confusion related to the terminology of emerging LCIA methods for biodiversity loss, water cycle disturbances and other impact categories. Clarity in terminology will improve the clarity of discussions on the conceptual and modelling challenges of assessing impacts. Eventually, the LCA community should try to build a consensus on both the terminology and the methods and practices for assessing these impacts. In the case of water cycle disturbances, this would require the harmonisation of on-going consensus processes for water footprint methodology (more or less intended for use in LCAs), including the ISO 14046 standard on water footprinting (ISO 2014), and the work by the WFN (Hoekstra et al. 2011), the UNEP-SETAC life cycle initiative (WULCA 2014) and the World Business Council for Sustainable Development (WBCSD 2015).

For further developing the theory of improved role selection in different kinds of technical R&D projects, it will be important that LCA practitioners share their experiences from various contexts. This goes against the norm of publishing only successful outcomes. Also, there is a need to test in practice whether an evaluation of the four project characteristics identified by Sandin et al. (2014) actually leads to more context-aligned role selection and improved use of the LCA work in inter-organisational R&D projects. In particular, testing the theory in contexts where other LCA roles are selected could help in developing the theory further—for example in terms of modifying the proposed project characteristics or adding other ones—and expanding its applicability.

To conclude, it is apparent that many aspects of LCA of forest products deserve further research. Meanwhile, it is important to acknowledge that LCA is not the only tool for dealing with environmental concerns of forest products, as emphasised in Sect. 3.1.2. A good reminder of this is the words of Maslow (1966, p. 15): "I suppose it is tempting, if the only tool you have is a hammer, to treat everything as if it were a nail". In relation to this, there is need for research on how to, in a relevant way, use LCA and other environmental assessments tools within the same study. This includes critically evaluating previous attempts in this direction, for example as done by Harder et al. (2015) for LCA and QRA. Also, there are other dimensions of sustainability, such as human rights, working conditions, equality and other social sustainability concerns, which are better dealt with by tools such as SLCA (Benoît and Mazijn 2009).

References

Benoît C, Mazijn B (eds.) (2009) Guidelines for social life cycle assessment of products. http://www.unep.org/publications/search/pub_details_s.asp?ID=4102. Accessed Feb 2015

de Baan L, Alkemede R, Koellner T (2012) Land use impacts on biodiversity in LCA: a global approach. Int J Life Cycle Assess 18(6):1216–1230

De Souza DM, Teixeira RFM, Ostermann OP (2015) Assessing biodiversity loss due to land use with life cycle assessment: are we there yet? Glob Change Biol 21:32–47

EC (2010) International reference life cycle data system (ILCD) handbook—general guide for the life cycle assessment—detailed guidance. Joint Research—Centre Institute for Environment and Sustainability. Publications Office of the European Union, Luxembourg

Ekvall T, Azapagic A, Finnveden G, Rydberg T, Weidema BP, Zamagni A (2016) Attributional and consequential LCA in the ILCD handbook. Int J Life Cycle Assess 20:1–4

Harder R, Holmquist H, Molander S, Svanström M, Peters G (2015) Review of environmental assessment case studies featuring elements of risk assessment and life cycle assessment. Environ Sci Technol 49(22):13083–13093

Hoekstra AY, Chapagain AK, Aldaya MM, Mekonnen MM (2011) The water footprint assessment manual: setting the global standard. Water Footprint Network, Enschede

ISO (2014) 14046: Environmental management—water footprint—principles, requirements and guidelines. International Organisation for Standardisation

Koellner T, de Baan L, Beck T, Brandão M, Civit B, Goedkoop M et al (2013) UNEP-SETAC guideline on global land use impact assessment on biodiversity and ecosystem services in LCA. Int J Life Cycle Assess 18(6):1188–1202

Lindqvist M, Palme U, Lindner JP (2015) A comparison of two different biodiversity assessment methods in LCA—a case study of Swedish spruce forest. Int J Life Cycle Assess. doi:10.1007/s11367-015-1012-6

Maslow AH (1966) The psychology of science. http://www.scribd.com/doc/133355995/Abraham-Maslow-Psychology-of-Science-pdf#scribd. Accessed Feb 2015

Røyne F, Peñaloza D, Sandin G, Berlin J, Svanström M (2016) Climate impact assessment in life cycle assessments of forest products: implications of method choice for results and decision-making. J Clean Prod 116:90–99

Sandin G, Peters GM, Svanström M (2013) Moving down the cause-effect chain of water and land use impacts: an LCA case study of textile fibres. Resour Conserv Recycl 17:104–113

Sandin G, Clancy G, Heimersson S, Peters GM, Svanström M, ten Hoeve M (2014) Making the most of LCA in inter-organisational R&D projects. J Clean Prod 70:97–104

WBCSD (2015) The WBCSD global water tool. http://www.wbcsd.org/work-program/sector-projects/water/global-water-tool.aspx. Accessed Feb 2015

WULCA (2014) Consensual indicator project. http://www.wulca-waterlca.org/project.html. Accessed Jan 2016

Zamagni A, Guinée J, Heijungs R, Masoni P, Raggi A (2012) Lights and shadows in consequential LCA. Int J Life Cycle Assess 17:904–918

Chapter 6
Concluding Remarks

Abstract This chapter contains some concluding remarks related to the content of previous chapters, for example stressing the need for context-aligned methods and practices in life cycle assessment (LCA) of forest products.

Keywords Context adaptation · Communication · Intended audience · Consciousness

A key message of this book is the importance of context-aligned methods and practices in LCAs of forest products; for example, in terms of selecting an appropriate role of LCA in a certain context (e.g. in an R&D project) and planning the LCA work in agreement with this role, modelling scenarios for capturing important product system uncertainties and methodological uncertainties, handling multi-functionality, selecting LCIA methods, and interpreting results. This relates to the fact that the main value of LCA lies in the decision support it can provide in a specific context. In other words, "LCA can be defended as a rational tool only through its use in decision making, and not as a scientific measuring device" (Hertwich and Hammitt 2001, p. 265). This suggests there are limits to the potential for prescriptive standards for carrying out LCAs of forest products. In other words, LCA requires practitioners that are conscious about the opportunities and limitations of LCA, and who adapt practices to this knowledge. For sensible use of LCA results, also decision makers ought to have a certain degree of understanding of the opportunities and limitations of LCA, and have the capability to discern LCAs that do not fit the specific decision context, from high-quality LCAs which fit the context at hand. It is most often the responsibility of the LCA practitioner to ensure that the intended decision makers have this understanding, for example by clear and transparent communication (e.g. in terms of what the LCA results can and cannot tell us in a specific decision situation). Related to this, it should be stressed that the art of LCA is not only about modelling the product system and calculating LCIA results, it is as much about interpreting results in light of a specific decision context and about communicating this interpretation to the intended audience.

G. Sandin et al., *Life Cycle Assessment of Forest Products*,
Biobased Polymers, DOI 10.1007/978-3-319-44027-9_6

73

Hopefully, by introducing important challenges and providing examples of some solutions, this book can contribute to increasingly conscious and context-aligned methods and practices in LCAs of forest products. Besides, the interested reader can hopefully use the book as a gateway to further reading. The reference list was up-to-date in early 2016, but as the methods and practices of LCA are in continuous development and permanent change, LCA practitioners need to constantly keep in touch with new developments in the field. Also, it is important that as many LCA practitioners as possible contribute to improving the methods and practices of LCA, by sharing experiences and insights in the academic literature and elsewhere. Only in this way, LCA can continue to provide relevant and useful decision support in an ever-changing world.

Reference

Hertwich EG, Hammitt JK (2001) A decision-analytic framework for impact assessment. Part 2: midpoints, endpoints, and criteria for method development. Int J Life Cycle Assess 6(1):5–12

Printed in the United States
By Bookmasters